HANG TIME

HANG TIME

My Life in Basketball

ELGIN BAYLOR

with Alan Eisenstock

Houghton Mifflin Harcourt

Boston   New York

2018

18 Apr11 11
B+T
27.00 (14.72

For information about permission to reproduce selections from this book,
write to trade.permissions@hmhco.com or to Permissions, Houghton Mifflin Harcourt
Publishing Company, 3 Park Avenue, 19th Floor, New York, New York 10016.

hmhco.com

Library of Congress Cataloging-in-Publication Data
Names: Baylor, Elgin, author. | Eisenstock, Alan, author.
Title: Hang time : my life in basketball / Elgin Baylor with Alan Eisenstock.
Description: Boston : Houghton Mifflin Harcourt, [2018] | Includes index.
Identifiers: LCCN 2017057002 (print) | LCCN 2017045607 (ebook) |
ISBN 9780544618459 (ebook) | ISBN 9780544617056 (hardback)
ISBN 9781328573582 (signed edition)
Subjects: LCSH: Baylor, Elgin. | Basketball players—United
States—Biography. | African American basketball players—Biography. |
BISAC: BIOGRAPHY & AUTOBIOGRAPHY / Sports. |
SPORTS & RECREATION / Basketball.
Classification: LCC GV884.3.B39 (print) | LCC GV884.3.B39 .A3 2018 (ebook) |
DDC 796.323092 [B]—dc23
LC record available at https://lccn.loc.gov/2017057002

Book design by Faith Rittenberg

Printed in the United States of America
DOC 10 9 8 7 6 5 4 3 2 1

To Elaine
You mean everything to me

CONTENTS

Reach up your hand . . . and take a star.

— LANGSTON HUGHES

1

GREEN WAVE

FEBRUARY 3, 2016.

I'm flying. Heading home.

I'm going to D.C., where I grew up, to visit my family and retrace my roots, maybe for the last time in my life. I'm eighty-one, and I haven't been back in more than twenty years. I'm not sure when I'll go back again. *If* I'll go back again. I hesitated to come on this trip, if you want to know the truth. I didn't know if I wanted to dredge up a lot of memories, to relive my time in D.C. Don't get me wrong: I have a lot of good memories—time with my family and friends, and mostly, of course, playing basketball. But the District was a different place back then. A hard place. A racist place. Segregated parks, schools, movie theaters, lunch counters. I had run-ins with the police. I experienced ugly, unforgettable things. One event in particular changed my life. I told myself that once I left D.C., other than to visit my family, I wouldn't go back. A lot of people who grew up in D.C. at the same time as I did feel the same way. They love the people; they don't love the city. Something about it makes you uneasy. You're always looking over your shoulder. The only place I ever felt totally comfortable was on a basketball court. That was home.

One time, I was walking from my house to school, Giddings Elementary. I was eight or nine. All of a sudden, a hawk swooped out of the sky and snatched a rat that had darted right in front of me on the sidewalk. Came right down, *whoosh,* grabbed the rat in its talons, and flew away.

That's the D.C. I knew: Rat City.

Now, flying east, heading home, I feel something stirring inside me, a stab of memory, and I find myself suddenly yearning to take a look into my past, if only for these few days. One last look.

The captain announces that we've reached our cruising altitude, and I settle back in my seat. I tilt to my right and continue listening to the man across the aisle, the man who's talking to me. A man I know well: Jerry West.

I had no idea he'd be on this flight and sitting across from me, aisle seat to aisle seat, nearly elbow to elbow. Jerry, who works for the Golden State Warriors, is flying to D.C. because the Warriors will be playing the Washington Wizards. The next day he'll go to the White House to meet President Obama, who will honor the 2015 NBA champions. That's become an annual tradition. You win an NBA championship, you get invited to the White House. At the moment, Jerry is talking about Steph Curry, the Warriors' star, the league's MVP, the greatest shooter in the world, the best shooter I've ever seen. I played with the second best—the man I'm talking to across the aisle.

"Oh, he's *great,*" Jerry says. "Except for the turnovers—"

I have to smile. Leave it to Jerry, an all-world perfectionist, to bring up Curry's one flaw. I don't feel like mentioning that I, too, will be going to the White House to visit President Obama. I'm not going with any team—I've never been on an NBA championship team—but I will be having a private visit with the president, just me and my wife, Elaine, who set it up.

Jerry and I talk for a while, then he reaches for a magazine in the seat pocket in front of him, rolls it up, and says, "We had some team, Elg, didn't we?"

"We did."

"We came close, what, how many times, *seven?*"

I nod. "We couldn't match up with Russell."

Jerry pauses, then rests his hand on my arm and whispers with an urgency I don't expect. "You should have a statue."

"What?"

"At Staples. There should be a statue of you. Believe me, I have asked . . ." His voice trails off. "I'm going to keep on asking."

I don't say anything. I picture the Star Plaza at Staples Center, a cluster of eight life-size bronze statues of Los Angeles sports legends that greet everyone who enters the arena—Wayne Gretzky, Oscar de la Hoya, Magic Johnson, Chick Hearn, Luc Robitaille, Kareem Abdul-Jabbar, Shaquille O'Neal, and Jerry West. All deserving. All L.A. heroes. Funny, though—I go further back than all of them. I was a *Minneapolis* Laker when the team was struggling to stay afloat. The owner, Bob Short, drafted me number one and later said I saved the franchise. He told me that I made it possible for him to move the team to Los Angeles.

I laugh to myself. I was a Laker when the name actually made sense. Minnesota, Land of Ten Thousand Lakes. L.A. is known for a lot of things, but *lakes* are not one of them.

Jerry, it's all right, I think. *I don't need a statue.*

• • •

February 5, 2016, 2:30 p.m.

Sunlight bounces off a snowbank and causes me to squint. I shiver underneath the collar of my leather jacket, shove my fists into my pockets, and follow three D.C. police officers as we crunch across the snowy blacktop to the back door of the boarded-up building.

Spingarn High School, my old school.

One of the officers fumbles with some keys, shoves the door open with his shoulder. I take a breath, duck my head, and enter the dark-as-night hallway. The police lead the way with flashlights as I stumble behind them, stepping through a layer of rubble. In the hazy light I see that we're walking through mounds of plaster and torn slabs of drywall. I step slowly, carefully. The air smells of smoke and dry rot. I feel as if we're edging down the hallway of a bombed-out building. The officer in the lead shouts back, "Watch your step, Mr. Baylor."

We go another few feet and turn right. Two officers pull open a set of double doors, pale light shimmering behind them. I lower my head and walk into the gymnasium, where more than sixty years ago I played forward for the Spingarn Green Wave.

"It looks ... smaller," I say, blinking into the funnels of daylight that pour through a half-dozen high, barred windows. The blond wood floor looks surprisingly shiny: not exactly polished, but in good condition. I start to walk toward the far basket, the floor squeaking beneath my loafers. I peer down the length of the gym. The light falls in orbs before me, splashing onto the floor. The effect is almost celestial. In comparison with the wreckage of the hallway, this gym feels like a cathedral.

"You could almost still play in here," I say, and the people around me laugh.

I keep walking toward the far basket. "Glass backboards," I say at half-court. "We didn't have them. We had fans."

I take a few more steps, reach the top of the key ... and my mind plays tricks on me ... messes with me ... because, ridiculously, I hear—

A crowd cheering.

A deafening roar.

Kids screaming.

Feet stomping.

A chorus of voices rising, chanting, "Rabbit, RABBIT, *RABBIT* ..."

I let the memory dissolve, the way I used to mute all sound around me. When I played, I blocked everything out—all noise, all distracting motion—focusing all my attention on just the game: the man guarding me, my teammates, the ball, the rim, this play, this moment.

I played in silence. I played without thought. I played by instinct. I played with complete concentration.

Sixty-three points in one game. D.C. high school scoring record. That record still stands.

I step up to the free throw line and look to my right. Unfurled from one of the barred windows, a banner flaps slightly. I make out my name and my number — 22 — and beneath those, a list of accomplishments scrolls down —

FIRST TEAM ALL MET

D.C. AREA BEST BASKETBALL PLAYER 1954

LEAGUE SCORING AVERAGE: 36.1

COLLEGE ALL-AMERICA 1957–1958

NBA #1 DRAFT PICK 1958

NBA ROOKIE OF THE YEAR 1958–59

10 TIME ALL NBA FIRST TEAM

11 TIME NBA ALL STAR

27.4 POINTS PER GAME

13.5 REBOUNDS PER GAME

ELECTED NBA HALL OF FAME 1977

"Elgin?" Elaine arrives behind me and gently rubs my back. "What are you thinking about?"

"Nothing," I say, my eyes riveted on that banner, my thoughts jumbled, memories pummeling me now, overwhelming me. "I just—"

"I know," Elaine says. "It's a lot."

"Yeah," I say, drifting to below the backboard. I reach up and brush my fingers against the rough twine of the net.

"It is," I say, and for a moment I picture myself back on the plane, talking to Jerry West.

Jerry, I'm not asking for anything. I never have. I'm happy with what I've got. Oh, I've gone through some things, like everybody else, but I'm a survivor and I consider myself very lucky. No: I consider myself blessed.

But I guess I wouldn't mind that statue.

2

RABBIT

DECEMBER 1933, CAROLINE COUNTY, VIRGINIA. DEEP COUNTRY.

Uzziel, my mother, John Baylor, my father, and their four children live on a farm, growing vegetables, caring for their few animals, and working their small plot of land. In addition to helping with the farm, Mother spends days at a time in the hospital with John Levi, her five-year-old son, who suffers from violent asthma attacks. One night, after the children have gone to bed, she reveals some disturbing news to her husband.

"I'm pregnant," she says.

My father, a stern and extremely quiet man under most circumstances, says nothing, knowing that, particularly in this instance, he really has nothing to say.

"How in the world can I do this?" my mother says, feeling overwhelmed, bordering on panicked. "It's too much. I cannot handle another child."

My mother decides to get an abortion. Someone tells her about an elderly white doctor out in the country who performs the procedure, and Mother makes an appointment to see him. My father drives her to the doctor's home office on a Saturday. He waits in the car while my mother explains to the doctor that she wants to terminate the pregnancy. The doctor says he's willing to perform the abortion but insists that my mother first consider all the implications and all the risks. My mother leaves, confused. On the way home, she changes her mind. She decides to have the baby.

The next day after church, as my mother leaves the small sanctuary, a lady she's seen at services approaches her. Mother smiles at the woman and the woman smiles back. They have never spoken to each other before. The woman's smile brightens, and then she slowly waves her hand in front of my mother's stomach.

"Don't worry about this baby you're carrying," the woman says.

My mother's jaw drops. She hasn't begun showing yet and has told no one except her husband and the doctor that she is pregnant.

The lady continues to move her hand back and forth in front of my mother's belly, and then she closes her eyes and nods. "This child," she says. "This child is going to be a blessing to you and your family."

My mother begins to cry.

Nine months later, on September 16, 1934, I am born.

· · ·

While my mother is giving birth, my father comes up with my name as he glances at his Elgin watch in the waiting room. He likes the way the word comes off his tongue. *Elgin*. It sounds regal. Plus he knows his Elgin watch to be steady, dependable, and on time, all admirable qualities that he hopes to pass on to his new son.

When I am nine days old, Mother and Pops move the family to D.C. Pops works two jobs: days as a custodian in a high school, nights driving a cab. Mother works as a maid in a hotel and then at one point goes back to school and earns a credential to become a secretary. She lands a job at the Department of the Interior, where she will work until she retires, more than thirty years later.

While Mother and Pops work, my sister Gladys, the oldest of us — eight years older than me — acts as a second mother, cooking, cleaning, and minding us. My other siblings all call her "Grandma," a nickname that spreads and sticks even among her friends, although I always call her just Gladys.

After Gladys, my siblings arrived in approximate two-year intervals: Arthur Kerman, whom everyone calls "Kermit"; John Levi, whom people call Levi or—I have no idea why—"Sal"; and then my sister Columbia, whom everybody but me calls "Fox."

From the moment Mother brings me home from the hospital, I have a special bond with Columbia. Not yet two, Columbia trails Mother like a shadow as she places me in my crib. She then stands on her tiptoes, watching me sleep, her head not reaching the top of the crib. Columbia finds me fascinating, poking me in the tummy with her finger as she tries to get me to gurgle or smile. To her, I am more than her little brother; I am her new living, breathing doll. As I get older, she dresses me and combs my hair the way she wants it—undoing the way Mother has combed it. Columbia keeps me by her side, takes me outside with her to play. She becomes my companion and my protector.

· · ·

From the time I can walk, I run. I get into everything, chasing after my older brothers, wanting to do whatever they do, racing to keep up with them. Kerman and Sal are big for their age—Kerman will grow to six feet eight inches, Sal to six foot six—and they are into sports. They play whatever sport is in season: basketball, football, baseball. They play at the park around the corner from us or on the street in front of our house. We live in a small row house on Heckman Street, and soon we move to a bigger house across the street.

On summer evenings and weekends, my mother and my sister Gladys join our baseball game. Columbia stands to the side and watches, pretending to be aloof. While Gladys is a tomboy, Columbia considers herself glamorous and not into playing sports or running after boys —she would prefer to have them run after her. Mother is tall, graceful, and athletic, and people say I look exactly like her. She's also energetic and competitive. She swings the bat hard, makes solid contact—*whack* —hollers, and runs the makeshift bases we've set up. Pops never comes out to play with us. Once in a while he'll walk by on his way to his night shift, driving a cab, and pause for a second to scowl at us, or I'll catch him open our front door and stare at us, grim-faced, like a judge.

We practically live outside. Unless there is violent rain or a rare heavy snowstorm, we're on the street or at the park. When I'm younger, my tricycle is my transportation. I nearly burn rubber to keep up with my brothers. As I get older and outgrow my tricycle, I rely on my legs. We play tag, hide-and-seek, and games like war and army. I'm *fast,* the fastest kid around, even faster than the older kids, and I know only one speed: full. One time, running through a neighbor's yard, I roar around a corner and run right into a clothesline. The rope literally lifts me off my feet. For a moment I lie on the ground, stunned, my neck burning. I sit up and hear laughing. Columbia stands a few feet away, her hands on her hips, shaking her head. She helps me up, still chuckling, but I can tell she's not laughing because she thinks my running into the clothesline was funny — she's laughing because she's relieved I'm all right.

When I'm around eight, Columbia takes me to a small local carnival. Somehow we get separated. As the crowd pushes and jostles me and I try to find her, I feel more confused than panicked. I walk around looking for her, keeping cool, knowing she can't have gone far. I've never seen so many people crammed together in one space, and I start to feel nervous.

"You look lost."

I peer up. A white guy around Kerman's age — sixteen, maybe older — stands next to me.

"I'm looking for my sister," I say. "Columbia."

"Yeah, I know," the guy says. "She told me to get you. We're gonna drive you over to her. Come on."

He grabs me by the arm and leads me to a car idling at the curb. Inside are three other guys. I don't know any of them.

I know that something's not right, but I allow the guy to drag me into the car. As we pull away from the curb, I peek out the window. There, desperately looking in every direction, calling my name, is Columbia. The car picks up speed. I open the door and jump out. I roll onto the ground as the car skids to a stop ahead of me. The car door flies open and the guy who grabbed me bursts out. I spring to my feet. The guy shouts something at me and starts to chase me.

I run.

He comes after me, closes the gap, and reaches for me. I spin away, duck, and run faster. I veer left and hit a higher gear. The guy, who's much older and bigger than I am, grunts and runs harder. I find yet another gear. I cut sharply to my right and then I bob, weave, stop on a dime, suddenly accelerate, and burn away from this guy.

I'm like . . . a rabbit.

I look back. The guy has stopped. He's bent over, panting, his palms on his knees, his face crimson.

I run even faster then, just because I can.

A while later, I look back again. The guy and the car have gone. I slow to a walk, head into the crowd at the carnival, and find Columbia. She screams at me for letting go of her hand. I promise that I will never leave her side again — ever. She clasps both her hands over mine and we start back to Heckman Street.

Rabbit, I think. *Rabbit.*

That becomes my name. That becomes my identity.

· · ·

I go to Giddings Elementary, an all-black school up the street and around the corner, a five-minute walk from my house. To get to Giddings, I walk past an all-white elementary school and then, if I want to save time, I cut through the park, the one on the left — the black park, which faces the white park. It's easy to tell them apart. The white park has a basketball court, baseball diamond, football field with goalposts, tennis court, and swimming pool, as well as a playground with swings, slides, and a climbing structure, park benches, neatly trimmed grass, and freshly planted flowers.

The black park has a sandbox and a swing. Nothing else. No facilities, no benches, no basketball hoop. The grass runs wild. The black park is one step up from a field.

But even as Columbia and I walk to school, past the white school and the white park with all the facilities, we don't complain, we don't question. The year is 1943, and this is how it is. We believe this is how it's supposed to be. How *we're* supposed to be. Separate. We mind our own business. We don't want to rock the boat. We don't want to cause any trouble or, worse, *be* the cause of any trouble.

Police officers in cruisers and on motorcycles watch us. White cops. I've heard people call D.C. "Chocolate City" because of the large number of black police officers who work the District. They must be working some other section, because in all my years on Heckman Street I never see one black police officer.

The cops circle around the black park and the white park and then go up and down our street. If they see more than two black kids standing in front of the store at the corner of our street, they slow down, stare at us, drive away, and then circle back. They never say anything. We have no relationship with them. They are faceless to us, and I know we are faceless to them.

We all fear them. We hear stories about random arrests and people getting beaten for no reason. I hear about a guy who gets picked up by the cops and disappears. Nobody ever sees him again. One night, someone breaks into a house on our street and the cops appear. They circle once, twice, circle back, and then pull over and round up a few kids, including my brothers. They shove Kerman and Sal into the back seat of their cruiser and haul them off to jail. My brothers are guilty of nothing except being two tall black teenagers talking in front of a store. My mother goes to the police station and the cops release them the same day. But now my brothers are on a list, because every time someone reports a crime in our neighborhood, the police arrive and take them away for questioning.

One time, after another break-in, I'm standing with my brothers in front of the store. We wonder what was stolen and who might be responsible. Suddenly a police car screeches to a stop, trapping us between the curb and the storefront. Two cops rush out of the car and herd my brothers and me toward the back seat. We fit the description of the suspects, they say.

In other words, we're black.

I dare not speak. Even if I want to speak, I can't. I am frozen with fear.

"Take your hands off him!"

Mother. Running toward us. She lopes down the street, erasing the distance between her and the cops in three graceful strides. My mother, a beauty when her face is at rest, has the look of a lioness when she is upset. And she is *upset*. She is outraged. "He's nine years old!" she shouts, and then goes into a tirade. I can't recall her exact words, but within seconds I'm standing next to her, her arm around my shoulders, her hand shaking as she watches the cops drive away with Kerman and Sal for what seems like the hundredth time. My mother returns me to the safety of our living room, where Pops looks through the window at the police car driving down the street with his two sons. I don't understand why my mother had to come to my rescue instead of my father, but I will soon.

My father, too, is intimidated by the police. They wear badges. They carry nightsticks. They carry guns. Growing up in rural Virginia, in former slave country, my father heard stories of how white cops arrested black people for anything, for nothing, then charged them with nonexistent crimes and assigned them to "work gangs," code for modern slavery. My father—quiet, proud, angry, and yet made impotent by the mere notion of the police—stands silently in our living room, his fingers wrapped around a shot glass half-filled with Colonel Lee Kentucky Straight Bourbon Whiskey, his eyes clouded with hate and fear. He understands that he has only one choice, which is to do nothing, brutally aware that if he intervenes on behalf of his sons, the cops will take him in, too. And then interrogate him. Or worse.

. . .

I play basketball for the first time at night, under the flickering glow of a streetlamp, on the forbidden basketball court of the white park. We play with a tennis ball. Late one night, Pops out driving his cab, Mother passed out from exhaustion, I follow my brothers out the front door, up the street, past the corner store, and over to the chained and locked fence guarding the basketball court. Heads swiveling, keeping an eye out for cops, my brothers reveal a tunnel they've dug beneath a section of the fence. I know we're playing with fire, but at that moment I'm a nine-year-old outlaw and I'm more excited than nervous. I scramble into the hole behind my brothers and slither under the fence and onto the concrete basketball court.

We find the tennis ball, a stray, resting on the free throw line. Beneath the streetlamp light that dusts us like snowflakes, we romp, trying to guide the tennis ball through the hoop. My brothers play in a basketball league at a nearby recreation center, but I have never played the game before. My brothers play an inside game, two towers hanging out beneath the basket, and they play rough. I can't get past them —they're too tall and physical—so I have to use my speed and quickness to somehow maneuver around them. They're human trees, but I am Rabbit, darting this way and that, looking to create an opening in which to shoot. At first they swat away every shot I attempt, but that only makes me more determined. I realize, too, that, like a rabbit, I have hops. This Rabbit can *jump*. Even though my brothers have six inches on me or more, I make them work. That game dissolves into a joyful free-for-all version of keep-away and what some call "army basketball," every man for himself.

We don't press our luck by playing too long, but this becomes a ritual. We sneak onto the court nearly every night, bringing friends, including a guy I'll call Sneaky Pete, who has been blessed with a very singular and useful talent: he can get you anything. At various times Sneaky Pete steals candy bars, cigarettes, food, liquor, steaks, and, when we're older, a car, which he hot-wires and I drive to the beach for the day. Don't know how or where he gets this stuff. Don't want to know.

One night Sneaky Pete shows up at the white park with a volleyball.

It's slightly deflated, but it's better than the tennis ball. Then, finally, one night he arrives with an actual basketball—mine to keep. We greet him like a hero returning with a bag of gold. We make teams and play the mid-1940s version of the game—a lot of hard-nosed inside play punctuated by an occasional, usually errant two-hand set shot. Then everything changes.

Walking home from school one afternoon, I see a vision. In the black park, a lone basketball hoop sways slightly in the brisk autumn wind. It has a rim, no net, and a wobbly wooden backboard, but it's there, majestic, inviting, all ours. No. All mine.

. . .

Basketball becomes my obsession. I play at the park after school, on weekends, and, once the school year ends, almost every day during the stifling D.C. summer, until it gets too dark to see the rim. My basketball is my sidekick, my partner. When I'm not dribbling the ball back and forth to the park, I keep it cradled against my side, like an appendage of my hip. At home, if I'm not around, my family never asks, "Where's Elgin?" Everyone knows the answer: "At the park."

On summer nights, without benefit of air-conditioning our small house becomes as oppressive as a sweatbox. My brothers and I often sleep in the park, spreading blankets on the ground. We know we're supposed to leave the park at sundown, but sometimes we chance it, staying through the entire night even though we're technically breaking some law about loitering. Other times, at two or three in the morning, a deafening metallic roar will snap us awake and a cop on a motorcycle will roll down the street and drive us out of the park, scattering us like mice.

In my family, I'm closer to the women than I am to the men. My brothers are so much older and live in their own world, a world they keep private, often behind closed doors, even though we share a room. They plot their futures. Kerman, about to turn eighteen, plans to join the military. Sal has his own ideas and talks about going to college. And, frankly, I try to avoid Pops. I especially try to stay away from his bad side. Pops doesn't say much. His face seems frozen in a permanent scowl. One of my sisters calls him "the original angry black man." I can see that. On the floor, by the side of the overstuffed armchair he always occupies when he's home, within easy reach of his right hand, he keeps a weathered leather strap.

So far I have avoided the strap, even though I probably deserve it as much as, if not more than, anyone else in the house. I have been spared Pops's giving me any "whuppings" because Mother always intervenes and literally saves my hide. I still live in fear of the strap, although I do wonder sometimes if it's really more of a threat than an actual instrument of torture. Pops can usually get his point across—his point being *Stop what you're doing*—just by lowering his newspaper from in front of his face and glowering. I know that parents shouldn't have favorites, but I can tell Mother considers me special. She calls me a blessing, which for me turns out to be a mixed blessing. She does save me from the strap, but this results in my brothers pounding the hell out of "Mama's favorite" when we're on the basketball court.

At home with Mother and "Grandma"—Gladys—I learn two important skills that will serve me well my entire life: how to cook and how to play cards. They teach me the joys of preparing a meal, how to bread, broil, and fry chicken and fish, how to chop and cook vegetables, and how to prepare a salad. I become their sous-chef and, by choice, their one-man cleanup crew. To this day, I can't stand the sight of a dirty dish. If I see a pile of dirty dishes, I roll up my sleeves and start scrubbing, even if I'm a dinner guest at someone's house.

On Sundays after church, and many Saturday nights, Mother entertains a group of ladies in our kitchen. They eat desserts, usually homemade pies, and play cards. I set myself up at Mother's elbow and watch, learning how to play whist, bridge, gin, and my favorite, poker. These women play for money and they play to win, which by now I've learned is the only way to play anything.

· · ·

Often during summers, when the heat invades our small house and sti-
fles us, Columbia and I escape. At thirteen, she has fallen in love with
the movies. We go to watch whatever film our local theater is showing,
walking the short distance from our house. Columbia refuses to see a
movie once. She will sit through each movie at least twice, sometimes
three times in a row. I'm no movie buff, but I do love air-conditioning, so I go with her without complaint. Usually, by the time the end
credits roll the first time, I feel my eyelids droop and I fall asleep. I sleep
fitfully—the threadbare seats actually feel less comfortable than the
ground at the park—and I begin a whisper chant that I want to go
home.

"*Shh,*" Columbia says. "In a minute."

By the time we leave the theater, it's usually past ten and I'm too
exhausted to walk, so Columbia carries me piggyback all the way home.
Even when I turn eleven and I'm tall enough that my feet drag on the
sidewalk, she still carries me on her back. She never complains.

One afternoon, Columbia picks me up after school and walks me
home. I try to jump on her back for a piggyback ride, but she ducks and
I slide off. We both break up laughing. I try climbing up on her again,
but she dodges out of the way and runs backward toward the edge of the
park. She senses someone standing behind her and stops. She turns and
comes face to face with a white girl her age.

"Nigger," the white girl says, and spits at Columbia.

Columbia stares at her, and then she slaps the white girl across the
face.

Columbia strides over to me, grabs my hand, and says softly, "Let's
go." We start walking through our park, heading home. In my peripheral vision, I see the white girl staring at us. I can almost feel the heat of
her hatred.

An hour or so later, my father sits in his armchair, thumbing through
his newspaper, sipping his precious Colonel Lee, while Columbia and I
sit on the living room floor doing homework. Suddenly someone raps
on our front door. Before any of us can move, the knock comes again,
harder, and a loud, impatient voice shouts, "Police! Open the door!"

Dread seeps into the room like a mist. My father puts down his drink, pushes himself out of his chair, and trudges heavily to the door. He opens the door to two young, heavyset white D.C. police officers. One looks past my father and scans the room, his gaze stopping on Columbia. The other cop stands framed in the doorway, shifting his weight, his hand resting on his gun.

"That your daughter, boy?" the first cop asks my father.

Boy.

The word slices into me.

My father—older than both cops—dips his head. He swallows. He doesn't seem to be able to speak.

"I asked you a question, boy," the cop says. "Is that your daughter?"

My father nods, his head down, his eyes boring into the floor, refusing to meet the cop's eyes.

"She's under arrest."

My father jerks his head up. He blinks in confusion. "For—what?"

"Assault."

"Assault?"

"She attacked a young girl in the park without provocation. She beat her."

Columbia rises to her feet and practically sprints to my father's side. "She spit on me and called me a n— . . . a *name*."

"You beat her," the cop says again.

"I slapped her," Columbia says. "But she—"

"That's assault," the cop says. "You're going to jail."

My sister starts to shake. My father swallows again and looks directly at the first cop, the spokesman. "Officer, sir, please, don't take her to jail—"

"Then punish her."

Again, my father blinks in confusion. "I will. She won't do it again. I'll see to it that—"

"Now," the cop says.

"What?" my father says.

"Punish her now."

The cop points a leather-gloved finger at something he sees in the room. An object at the base of my father's armchair.

The strap.

"Whip her," the cop says.

Time seems to stop. I want to run. Bile rises into my throat. I feel sick. I'm not conscious of moving, but I feel myself sliding backward on the floor until I'm nearly out of the room.

"*Now,*" the cop says again.

My father whimpers. "Yes, sir."

Columbia's entire body collapses and she begins to cry. I don't know what to do. I see the cops step all the way into our house and close the door behind them. I don't raise my head. I can no longer see their faces. I see only their boots.

I don't see what happens next.

I don't see because I don't look.

I can't look.

I don't close my eyes, but I narrow them to slits. I can see only blurry images in my peripheral vision.

I see my father grab Columbia roughly by the arm. I see her struggle, and then I see her wilt, her body going limp.

I see my father pick up the strap.

I see a brown blur, a whip flying in slow motion.

And then I do close my eyes.

What I hear is far worse than anything I would have seen.

I hear my father grunt.

I hear the thwack of the strap.

I hear Columbia's screams.

I try to shut out her screams. But I can't. I hear the strap. I hear the strap. I hear the strap . . .

I hear Columbia wailing like a wounded animal, then gasping, her sobs ripping through the room with pain and terror and humiliation.

Then silence.

I hear her footsteps, running.

I hear my father groan.

I hear the front door open and close.

I wait. I hold my breath.

I know that everything in my world has changed. I know that my sister will never be the same, that her soul has been damaged. I know that our relationship will never be the same. And I know that my relationship with my father will never be the same.

Boy.

Those two white . . . animals . . . called my father, a grown man twice their age, . . . *boy*.

He did nothing. He said nothing. He beat his own daughter in front of them while they watched.

I burn with anger. I feel so much anger, my stomach clenches.

And I feel *hatred*.

Raw hatred.

I have never felt hatred before.

I hate my father. I hate him so much.

I make myself a promise.

Someday I'm going to kill him.

I'm going to kill my father.

I tuck my knees into my chest and I sob.

• • •

I lose track of time. Months float by. I spend my days walking through a kind of darkness. I refuse to talk to my father. I won't even look at him. Columbia retreats into herself, speaking to no one, and then she begins wandering into a world outside that I don't know. She goes out alone. She comes back late. I'm afraid for her. I turn twelve, then thirteen, and I realize, sadly, that Columbia and I are coping with the trauma of that day by finding our own separate escape routes.

I escape into basketball. I don't go home after school. Instead I walk twenty minutes to the Southeast Settlement House, a recreation center that offers after-school activities, arts and crafts, music, and basketball. I join their team, even though I'm the youngest player by far. Southeast

Settlement House has no gym, so we travel, playing our games at other community centers in the area. I'm a benchwarmer, getting into games only for a few minutes at garbage time. But after I leave the Settlement House, I head to the playground and play pickup ball with kids my age. I'm taller than most of them and quicker and better than all of them. When it starts to get dark and I know my mother has gotten home from work and I won't be alone with my father, I head back to Heckman Street.

I turn fourteen and become a starter on the Southeast Settlement House team. Then I join the Southwest Boys Club team, because I crave tougher competition, and they have their own gym. The floor is uneven, the backboards have no give, and the lighting is dim, but at least we don't have to travel. I start to dominate in games, scoring pretty much whenever I want, especially inside near the basket, where I use my speed to drive by defenders who try to check me, or slip past defenders who set up underneath, trying to protect the hoop. I roar by them or soar above them, jumping higher and hanging in the air longer than they can, using English to bank soft shots off the fan-shaped backboards. I don't plan what I do. I just do it.

I try to shoot from outside the way everyone else does, with two hands over my head, but I find I have no control. I also don't like the idea of stopping, setting up, and then letting the ball fly from a stationary position. It feels awkward and inefficient and slows my flow. I start working on shooting with one hand while jumping at the same time. Almost immediately, this new "jump shot" feels natural, feels right. I have more control, better accuracy, and my jump shot allows me to shoot over anyone who guards me. I also feel a little different. Nobody I've seen shoots this way, although a few kids on the playground start to copy me. I don't care. I work on my jump shot from every spot on the court. I face the hoop, aim just over the front of the rim, and then I shoot from the sides, angling my shot off the backboard. I practice my shot for hours, sometimes on my own, most of the time in games —settling the ball onto my fingertips, arcing it, finessing it, perfecting it, making it mine.

Walking home from the park after a Sunday afternoon playing pickup ball, two friends and I stop at a corner house, drawn by a shaft of blue light coming through the front window. We crowd around the window and see that the light is coming from a television set in the center of the living room. Planted on a couch in front of the television, which is encased in a hulking wood cabinet, is a white family — mother, father, and a couple of kids. I know these people. The father owns the small market up the street, and I've seen the mother working the cash register.

We huddle closer to the window so we can see what they're watching. We make out a basketball game: our local team, the Washington Capitols, taking on an opponent I can't name. Two years ago, the Capitols became a charter member of the Basketball Association of America, the new professional basketball league. I follow the team religiously in the newspaper, reading the account of every game and studying every box score. I know all the Capitols' players, my favorite being their high-scoring forward Horace "Bones" McKinney. I've memorized his scoring and rebounding average — he leads the team in both categories — but until this moment, peering through this living room window at the small, blurry black-and-white images flickering on the television screen, I've never seen a professional basketball game. I try to pick out Bones McKinney, but of course I can't. I can only imagine.

Mesmerized, the three of us watch the game for most of the second half. It occurs to me that the family must know we're outside their house, watching the game through their window. While they won't, or can't, invite us in — that's not the way the world works in 1948 — they allow us to stay by their window and watch without running us off or closing the curtains. After that, my friends and I always go home by way of this house, stopping to stand outside and watch TV through their front window.

It won't be long before my father buys a television set, one that you have to operate by inserting coins in a slot on the side, like a vending machine. My mother works long hours, my brothers and my sister Gladys have moved out, Columbia keeps to herself, and my father and I don't really communicate. Pops plays music to fill the silence in the house, but the blues he listens to probably only brings out the sadness he carries. In my opinion, he bought that coin-operated television set for companionship.

Eventually, without saying much, I join him in the living room Friday and Saturday evenings to watch boxing and wrestling. I love boxing, as he does, but I don't really get the point of wrestling, which is more show than sport. My father totally immerses himself in the action, all of it choreographed and obviously fake. He edges forward in his armchair and shouts "Watch out!" to good guys like Bobo Brazil and Gorgeous George. And when a villain such as Dick the Bruiser or Killer Kowalski sucker-punches or karate-chops one of the heroes, he yells in frustration, "He deserved it."

One day the Southwest Boys Club holds a free-throw-shooting contest. Most guys shoot clunky two-hand set shots. A few shoot free throws underhanded. When it's my turn, I toe the free throw line, bend my knees, dribble once, and, using my one-hand technique, make every one of my shots and win the contest. A bigwig at the Boys Club presents me with my prize: a ticket to see the Washington Capitols.

I go to the game by myself. I settle into my seat and watch the players I've read about and seen only as shimmering figures on TV: Frank Scolari, Johnny Norlander, and of course Bones McKinney, a high-energy six-foot-six forward with slick moves around the basket. The Capitols dominate the other team. In 1947, the year before this, the Capitols opened the season by winning seventeen straight games, a record that will stand for decades, and now they seem even better. Their coach, a fiery young guy in a rumpled suit who prowls the sidelines rubbing his prematurely balding head—relaxing only when the Capitols clearly have the game won, and then, to the crowd's delight, sitting on the bench and lighting up a victory cigar—is Arnold "Red" Auerbach, who will become the best coach in NBA history when he goes on to work for the Boston Celtics, my greatest rival and my nemesis for more than a decade.

Back at the playground, I try out moves I saw Bones McKinney make and then I refine them, especially around the basket. I work on a head fake designed to get defenders to jump before I take a shot or drive to the basket. I'm too quick for the kids I play with, the ones my age. I itch to play against bigger kids, older kids, and—I hate to say it this way, but—better kids.

Even so, kids challenge me.

"Let's go one-on-one, Rabbit," one guy says. "Play you for a quarter."

I don't really want to play this guy. He's overaggressive, a butcher, and lacks skill. I don't want to get hurt. But a quarter does go a long way.

"All right," I say, flipping him the ball. "Take it out."

"Game's to 11," the kid says.

He tries a move and takes a shot. I block it and grab the ball before it goes out. My ball. I make eleven straight baskets and win 11–0.

"Let's run it again," the kid says, reaching into his pocket and handing me a quarter.

"No, man, I really don't want—"

"My father's a bank teller. I got a whole roll of quarters."

Well, in that case.

I beat this kid like a drum. I take enough quarters from him to buy myself lunch for a month.

I look for tougher competition. I meet a couple of guys who can play, but nobody can match me. I'm not bragging; it's just true.

I become friends with a tough player named Gary Mays. Everybody calls him "the Bandit" because he only has one arm, like a slot machine. I don't ask him how he lost his arm, but I hear he accidentally blew it off with a shotgun. Stocky, strong, built like a weight lifter, Gary plays every sport, including, unbelievably, baseball. What's even more unbelievable is that he plays *catcher*. Squatting behind home plate, his right hand inside his mitt, Gary catches the ball from the pitcher, flips the ball in the air, slides off his mitt, and, with his one hand, snags the ball and guns it back to the pitcher in one fluid, blazingly fast motion.

When we play basketball, the Bandit likes to be on defense. He gets right up on you, punishing you with his thick chest and on hot days whacking you with his sweaty nub, perspiration flying off it and flicking into your face. I can't stand Gary guarding me, because I hate dealing with that nub. He brags that he's the only one who can stop me from scoring. It's halfway true: I can score against him anytime I want; I just don't like getting anywhere near his nub.

One day I meet my match, a guy who lives on the next street over: Clarence Hanford. I find him at the park, alone, shooting around. He's taller than me and older. Much older. He plays for his college team. I ask him if he wants to go one-on-one. He tucks the ball against his hip and studies me. "How old are you?"

"Fourteen."

"Nah."

"Almost fifteen. In a month. Couple of months."

"All right. We'll play one game."

We go at each other. Clarence has good moves and plays tight defense. He doesn't concede anything. Still, I stay with him. We match basket for basket, but then he gets hot, hits three shots in a row, and beats me 11–8. We go again and again. I stay right with him. I try a head fake. He falls for it but then recovers and blocks my shot from behind. I grab the ball, dribble outside, hesitate, wiggle my shoulder, fake left, and blow by him on the right for a layup. Playing Clarence, nothing comes easy. We both fight for each basket. He wins the second game. Panting, sweating, we shake hands. We promise to meet at the park again for another rematch. And so we do, playing one-on-one, day after day. I battle him, keeping every game close — and I lose every time. Somehow I don't mind: playing with Clarence, I feel like I'm in basketball school, learning to compete against someone who's bigger, stronger, and older. Even though I keep losing, I'm living the old saying: If you want to improve your game, play against somebody better.

And then, after months of having Clarence kick my butt, I arrive at the park and notice that he and I are almost the same height.

"You've grown," he says.

"Maybe."

"Loser's outs," he says, handing me the ball.

I take the ball out and I score. Then I score again. And again. I win, easily. We run it back and I win again. We play a third time. Not only do I win, but Clarence doesn't even score. We play a few more times and I win every game.

Clarence never beats me again. We finally agree to quit playing against each other because we both know that he's no match for me. The guy I could never beat has become too easy to beat. I have only one choice now: I have to leave my neighborhood and find a tougher game.

• • •

After Giddings Elementary, I attend Randall Junior High, a thirty-minute walk from my house, where each day folds into the next . . . and the next . . . and nobody knows your name, sometimes not even the teachers. I feel like I'm stuck in a kind of holding pattern, biding my time —hell, wasting my time. I don't find the subjects boring—I particularly like history and English—but my future seems predetermined. I guess because the world sees black kids as capable only of manual labor, I'm headed for a vocational track. In ninth grade, I take courses such as machine shop, sheet metal shop, shoe shop, and automotive mechanics, courses I'll continue at Phelps Vocational High School, where I'll prepare myself for a career in automotive repair or, if I show an aptitude for plants and soil, a life of landscaping. Even though now would be the time to start prepping for college, the idea doesn't come up. I never receive any encouragement from the teachers. One of them—one of only two male teachers at the school—stops me on my way out of class to tell me that, in his opinion, I will never amount to anything. His words sting and then motivate me. *I'll show him*. I'm not going to end up fixing cars in some greasy garage. I have bigger plans. I'm going to be a gym teacher.

Once it's clear that I'm on the vocational track, I tend to drift in class.

I don't see the point in paying attention. I fantasize about basketball, focusing my thoughts on making the high school team. I'm tall, but I'm skinny, and I wonder if I'm good enough to make varsity.

Only one school subject holds my interest: music. I toy briefly with learning how to play the piano. My mother encourages me. I take a couple of lessons after school, but my buddies find out and make fun of me mercilessly. I quit taking piano lessons, but not because of them. I wouldn't mind *playing* the piano, but to do that I'd have to practice. And I hate to practice.

I do love listening to music, and especially to singers like Dinah Washington, Nat King Cole, and Joe Williams, the main singer for the Count Basie Orchestra. I even write down the lyrics to my favorite songs so I can sing along to the radio. In school, our music teacher, the funny and flamboyant Mr. Glass, makes us memorize a song and perform it in front of the class. Most of the kids, especially the boys, slouch to the front of the classroom and mumble something indecipherable into their shirts. Not me. I choose the old standard "Sylvia," and I belt it—I've been blessed with an outstanding singing voice, if I say so myself. Maybe I'll try to make it as a singer first, and if that doesn't work out, *then* I'll become a gym teacher.

• • •

I wait.

I wait as long as it takes.

I have my back to the basket. My man guards me close, both of his hands on me, his right hand leaning on my shoulder, his left hand pressing into my lower back, trying to push me out of the post. You could easily call a foul, but the ref, standing in front of us, doesn't care. He probably doesn't even see.

So I wait.

I flex my shoulder. The defender's right hand rolls off me. I dribble high, bounce the ball to my waist, showing him the ball, tantalizing him. He reaches for it; I dribble away. He whiffs, slaps the air. I allow myself to crack a smile.

I want to say, "You can't guard me. You know you can't guard me."

I snicker to myself.

Nobody can guard me. That's what I think, what I believe.

I would never say that aloud. I don't brag. I don't taunt. That's not how I was brought up. But I know it's true.

And now at this moment—less than a moment, really—I wait.

I want him to commit. I know he will.

I dribble high again, tease him with the ball.

Now I move . . . but . . .

I don't.

I fake.

He bites. He jumps.

I don't. Not yet.

I wait until he's coming down.

And *then* I jump. My back is still to the basket. My man's in the air.

He looks like he's flailing. I look like I'm flying.

And then I shoot over my shoulder, a casual flick on my way down. The ball kisses off the backboard. I hear the delicious tickle of the net.

The crowd gasps. That's what I'd call it—a collective gasp, as if coming from one large body. Usually I block out all sound, or try to. But this sound washes over me and I perk up. What I hear seems muted at first, distant, and then I hear what appears to be a . . . *sigh* . . . like air escaping from a tire. And then, as I run down the floor, getting back on defense, I hear a flood of voices chanting, "Rabbit! Rabbit! *Rabbit!*"

This stuns me, because we're two hundred miles away from D.C.

We're in Norfolk, Virginia. My team, Phelps Vocational, is playing Booker T. Washington in their season opener, on their home court.

We have no home court, because Phelps has no gym. Every game we play involves a bus ride, though most don't go as long as this one: four hours one way, from D.C. to Norfolk. We practice at a nearby junior high gym that looks smaller than regulation to me, and we play our few home games at a local arena when it's available. That's about to change, though, if I have my way.

A new high school, Spingarn, with a beautiful full-size gym, featuring floors so polished you could shave in them, opens this year, 1952. According to my mother, Heckman Street falls into the zone that the new high school will serve. I will be able to attend Spingarn for my senior year. Mother has seen me play a few times after work, but once I'm at Spingarn, she plans to come to every game, as will the rest of my siblings, the ones who are still around. I don't expect Pops to come to any of my games, though. He claims he can't take the time off from work. He also calls my basketball playing "foolishness." He wants me to buckle down and start thinking seriously about what I plan to do after high school, what kind of job I can get.

Mother has other ideas. We have a couple of conversations about basketball and my future. As Mother and I have talked, as we have dreamed, we've wondered if maybe the way I play basketball will attract the attention of some colleges. Especially if I transfer to Spingarn. Maybe.

The game in Norfolk ends. We annihilate Booker T. Washington, 78–53. I score 35. I have begun this year, my junior year, as our team's captain. Last year, my first on the team, I averaged 18.5 points a game. We dominated our league. You wouldn't know that, however, unless you came to our games or followed us in one of the area's black newspapers, such as *The Baltimore Afro-American,* because only the black papers cover our team. The so-called mainstream newspapers, such as *The Washington Post,* don't cover black high school teams. Every week, the *Post* ranks the top ten high school basketball teams in the area and they never mention us — or any black team. One of the higher-ups at Phelps decided that our team deserves more recognition, so this year, before we start league play, we hit the road, going on a sort of barnstorming tour, playing a half-dozen of the better all-black high schools in Virginia, including Booker T. Washington. Before Booker T. Washington, we played Peabody and Armstrong. We beat them all. I believe — no, I *know* — that we can beat any high school around, including any of the top white high schools. At least I'd like to play them and see how good they really are. I'd like to see ten white high schools that are better than us.

Because of basketball, I am becoming kind of *known*. Mother has cut out and kept every clipping mentioning me in the *Afro-American*, like the time I scored 42 against Cardozo, breaking a city scoring record. I'm proud that she's keeping a scrapbook, but I also feel embarrassed. I love playing basketball; I just don't love the spotlight. And even if I excel, my family keeps me humble.

The first time my sister Gladys sees me play, she's married and living in Harpers Ferry, West Virginia, a couple of hours away. She and her husband, Simiel, a pharmacist, drive up to see that game against Cardozo, stopping to pick up Columbia on the way. Afterward, Columbia, beaming, asks Gladys, "What did you think of little Elgin?"

"Show-off," Gladys says.

I laugh, but I receive her message loud as a shout.

Be humble.

I'll never forget that.

I play basketball my way, and I might even be a star on the court, but once the game ends, I'm the same as everybody else.

Still, it's hard not to notice me. I've reached almost my full height of six foot five, and I do play the game with a certain . . . flair. Some might call it flash. I don't see it that way. The way I play feels natural to me. I shoot unorthodox shots because the guys who guard me don't expect them. By improvising, I get open. There's nothing more effective than the element of surprise. You can't practice these shots, because they didn't exist before and they will never happen again. I don't plan them, I just *do* them, on the spur of the moment. Kind of like the way Miles and Coltrane play jazz: you never know what's coming.

Same with my passes. I like to snag the rebound, sprint downcourt, look one way, and pass another, to throw off the defense. These passes are no accident. I have this innate ability to *see* the whole floor and pick out my teammates out of the corner of my eye. I know exactly where I'm passing the ball. If I can get the ball to someone while he's running full stride toward the basket by throwing the ball behind my back or zip-ping it to him around my head, I will. Most coaches drill you to throw nothing but two-hand chest passes. That's fine; I can do that. But not if it slows me down and takes away my advantage.

When the high school season ends, the pickup season begins. I'm seventeen now, it's March, the last D.C. snow has melted, and after school and on weekends my friends and I play in the toughest playground games we can find. We go full-court, five-on-five, or sometimes half-court, running three-on-three. We play to twenty baskets, by twos, and we call our own fouls. You don't shoot foul shots; you just take the ball out. For some reason, the guys who lack skills, the hackers, commit the most fouls and call the most fouls. I rarely call a foul, unless I get clubbed over the head or some gorilla tackles me. My friends name me de facto captain, and I choose the guys for our team. We play "winners stay," meaning the winning team gets to keep on playing. If you lose, you sit and hope someone picks you to play the next game. We don't worry about that: we don't lose.

In addition to playground ball, I join the Stonewalls, a sort of club team made up of older guys. Some of the guys are out of high school and working, some are in college, some have graduated college, and some have come back from the army, like my brother Kerman, who's gotten married and taken a job working at a furniture store. These guys play rough. I can jump higher than anyone I play with, but I won't dunk or fly to the hoop unless I'm wide open, because these guys will undercut you. Some guys go for your legs. I take elbows to the face and forearms to the throat. I learn to stand my ground and body up. We play a few games in small gyms, but mostly we run outside on concrete courts. When we play outside, there will be blood. Guaranteed.

Between dominating during my junior season at Phelps and doing the same on the segregated playgrounds of D.C., I'm developing a kind of reputation. I feel like I'm riding the crest of a wave, and basketball has taken me there.

And then life, as I know it, ends.

3

LEGEND

IN MARCH, WHEN I'M SEVENTEEN, I MEET HER: BARBARA Arnold. The girl I will marry.

A friend's girlfriend introduces us at a party. She's a couple of years older and doesn't go to Phelps. She does, however, go to all my games. I guess these days you'd call her a groupie. I don't really notice any individual people in the stands when I play, but when Barbara and I meet, she tells me she has never missed a game. She calls herself my biggest fan.

Like every kid I know, I go to parties on weekends, house parties, where we dance and sometimes do more. The truth is, I don't do much more. I try to act cool, hang back, but since I'm always the tallest person in the room, I kind of stick out. And maybe because of my reputation, people think I'm experienced. I'm actually very inexperienced. I do know that girls check me out.

At one party, while I'm getting a drink, I overhear two girls talking about me. One says, "Now that's a long, tall drink of chocolate."

"One good-looking piece of beefcake," the other one says.

"I would like to climb that pole," the first one says.

"Excuse me." Columbia appears out of nowhere and pushes herself between the two girls. She eyes them both with menace. "You are talking about my *brother*."

The first one takes a step backward. "Oh, I didn't—"

"You both need to move—*now*."

The girls look at Columbia, decide they do not want to mess with her, and walk away, giving me a long glance as they leave.

Columbia takes my drink, sniffs it, realizes it's a soda, and hands it back. "You have to watch yourself," she says. "Girls will come on to you."

"I can't help it," I say, grinning at her. "Some guys got it, some guys don't."

"I'm serious, Elgin."

"Okay," I say. "I get it, Columbia," but I'm not sure I really do get it.

"Be careful," Columbia says. "You're a target."

• • •

April 1: the first time Barbara and I have sex.

I've talked to her a few times at parties. We flirted and then agreed to meet at her place. "We'll be alone," she said. Her mother would be at work and we'd have the whole house to ourselves.

I take the bus from school to her neighborhood and trudge up a hill to a cluster of small houses. I hesitate at her front door. Down the hill I see Spingarn, the sprawling new high school that will open its doors in September. I take a deep breath, ring the doorbell, and wait for Barbara. I feel uncomfortable. For a moment I consider turning around and running back down the hill. Then the door opens and Barbara smiles. She lowers her eyes shyly, takes me by the hand, and locks the door behind us. There's a finality in that—in her locking that door—that makes me shiver.

She leads me directly into her bedroom, and time seems to speed up. Everything that happens next vanishes into a haze—a flurry of furious motion, music playing, clothes flying, bedsprings singing, the room spinning. Barbara takes charge, because I feel so nervous and hesitant and unsure. At one point—although I'm embarrassed having this thought—I notice that Barbara is heavier than she appears when she's dressed, especially around the middle. When she's out in public, she covers herself shrewdly, wearing outfits that disguise her weight.

I close my eyes and try to disappear, but then I hear a noise. A distant click. I focus on that sound, try to identify it, and realize that I heard the front door being unlocked. Then I hear the door open and quick, urgent footsteps. Barbara's bedroom door swings open. A loud voice shakes the room, and Barbara's mother fills the doorway. She screams in a fury, and Barbara screams in response. I leap off Barbara's bed and I'm tripping and grabbing my clothes and pulling them on, hopping out of that room, thinking only, *I want to erase what just happened.*

I wish I could step into a time machine and turn back that afternoon. Then I remember Columbia's warning and I wish most of all that I had listened to her.

I bang out the front door, scramble down the hill, Spingarn High looming below, my breathing coming so fast, my heart thumping so hard I fear it will tear through my chest.

April 1.

April Fools' Day.

. . .

Summer 1952.

I lose myself in basketball. I live and breathe basketball. The only place I want to be is on a basketball court. I storm the District with the Stonewalls. I'm still the youngest player on the team, but I dominate, pouring in 30 or more points a game and grabbing 20 rebounds. We start to draw crowds. People pack the small gyms we play in and line up two or three deep at the playground courts. I mention this to Kerman. "A lot of people are coming out to watch us."

"No," Kerman says. "They're coming out to watch *you.*"

As for Barbara, ever since that afternoon, I've kept away from her. I chalk up what happened as a lapse in judgment, an unfortunate mistake. I got lost in the heat of the moment and allowed this older girl to take control of my mind and my desire. I promise myself I'll never let that happen again. I vow to heed Columbia's advice and be much more careful in the future.

For the first time, thanks to basketball, I've allowed myself to envision a future beyond high school, a future that doesn't involve working with my hands or teaching gym class. Over dinner, my mother and I talk about the possibility of my playing basketball in college. So far, no colleges have called, but we both believe that will change if I play well during the upcoming season. I start to fantasize about playing in college. I wonder if I'd be good enough. I wonder if I can hold my own at that level of play.

One evening, as I start to head home after a game with the Stonewalls, I see someone standing in the shadows under the basket at the far end of the playground. I approach slowly and make out who it is. Barbara. I haven't seen her in weeks.

"I need to talk to you," she says.

Barbara is the last person I want to talk to. I shrug and start to move past her. "What about?"

"I just came from the doctor." Barbara holds her breath for a moment and then exhales. "I'm pregnant."

I stop dead.

"You're . . . what?"

"You heard me."

I feel myself blinking rapidly. I start to speak, but I stammer uncontrollably and swallow my words. "It . . . it was one time—"

"Everybody knows you're a good shot, Elgin."

I stammer again. "Barbara . . ."

That's all I can say.

Barbara stares into me. "There wasn't anybody else. You're the father."

"Are you . . . ?"

"Don't even think about it. I'm having this baby." She starts to leave, then turns back and practically growls at me. "You better think about your responsibility."

"My—"

"*You owe me.*"

She hisses the words.

. . .

I don't know what to do.

I don't know who to talk to.

I don't even know what to think.

I'm too embarrassed to say anything to Columbia. She warned me. My brother Sal, whom I consider very rational, even spiritual, someone I could talk to, has joined the military. My sister Gladys has married and lives two hours away. I certainly can't talk to Pops. I can't talk to my mother—this news will kill her. I see Kerman almost every day, and I know he could at least relate. He also got a girl pregnant and ended up marrying her. I've never spoken to him about that: how it altered his entire life, and how he feels. I don't have to. I see how he feels. He's angry—angry as hell. He takes it out on the basketball court. He plays with a rage, ripping down rebounds and punishing anyone who guards him.

So I keep my feelings inside. I don't say a word. I choose not to talk to anyone.

I play basketball, come home, go to bed, and lie awake, staring at the ceiling, wondering what I should do, my thoughts jumbled as I desperately try to sort out the uncertainty ahead.

Before I know it, summer ends. Three months have seemingly evaporated.

Labor Day comes and goes. The first day of school arrives. My parents and I share a silent breakfast. They go off to work and I leave for school. Except I don't go. I wander around the neighborhood. I circle back toward my street and find myself at the park. I lower myself heavily onto a park bench, close my eyes, and lift my face toward the midmorning sun. Five minutes pass, then ten. I open my eyes. I know that I have to figure out what to do, and that I can no longer deal with this alone.

That night I tell my mother. I try to explain what happened and I can't speak. I stammer, the words caught in my throat. *I'm supposed to be my mother's blessing. What did I do?* I'm filled with shame.

Eventually the words spill out. My mother listens, quietly blinking back tears. When I'm finished, she sits up straight, almost regal. My mother always, even at this time of trial for me—for our entire family —somehow finds a reservoir of strength. She tells me that we need to have a family meeting. She says she wants no secrets. She says everything will work out for the best. I want to believe her. I have to believe her.

. . .

We gather in the kitchen, all of us squeezing around the table. Gladys and her husband, Simiel, sit at one end. Kerman and Columbia sit across from me. I sit between Pops and Mother. I start to tell everyone what happened, but Mother, my protector now more than ever, takes over. When she finishes speaking, my father exhales and slams his palm on the table. I jump. He kicks his chair back, gets up, and paces. He stops, rubs his head, and returns to his seat. "Well," he says, "you have to marry her."

"No," my mother says.

My father folds his hands in front of him, links his strong, thick fingers tightly. "He has to do the honorable thing."

My mother's voice rises. "He's too young. He's seventeen years old. He has his whole future to think about."

"He should've *been* thinking about that."

"*She* should've known better," my mother says. "She's older. I blame her."

Then, as if noticing me for the first time, my father speaks directly to me. He speaks flatly, without anger. "You got her pregnant. She's your responsibility."

I look down at the table. "That's the same thing she said."

"You're saying he should drop out of school, marry her, and get a job?" my mother says.

"Yes," my father says, and then looks at me, his eyes lasers. "You have to do what's right. You need to support this child."

"She didn't give him any choice," my mother says.

"He has no choice," my father says.

. . .

For several heated hours, my family discusses my future. Gladys speaks calmly, intelligently, her voice reasonable but tinged with sadness. Columbia spews, speaking with fire and passion. She places all the blame on Barbara, sees her as poisonous, a viper. "She did something nobody could do," Columbia says, bitterly. "She caught the Rabbit."

Through all the conversation, my father never wavers in his absolute conviction that I have to do the right thing—I have to marry Barbara. My mother argues with him. They shout, my mother cries and pleads her case, but in the end, one opinion prevails: my father's. I will marry Barbara.

Feeling more exhausted than after any game, I can do nothing but go along with the decision. My fate—my future—has been sealed. I'm seventeen and about to become a husband and a father. I will drop out of school and work for minimum wage alongside Kerman as a deliveryman at the furniture store.

A few weeks ago, a crowd gathered at a playground to watch me and the Stonewalls play a game of basketball. As we took our warm-up shots, Kerman turned his back on the court and silently counted the people. "There must be five hundred people here," he said.

"That can't be right," I said. "You can't fit five hundred people behind these sidelines."

"At least five hundred," Kerman said. "I think I'm low."

I was the show, I realized. These people had come to see me. Something was happening that seemed bigger than me. I was on the verge of entering a strange, unknown world, a world I couldn't wait to experience.

But now that world has been crushed, pulverized into dust.

That day — that game — seems like a lifetime ago.

A lifetime that belongs to somebody else.

. . .

I drop out of high school. The thought that not only have I dropped out but I have been forced to give up basketball paralyzes me. I can't seem to do anything. I can barely get out of bed. I can't eat, I can't sleep, I don't want to speak to anyone.

After a few days of this, Pops reads me the riot act. He tells me that I have to leave the house. He will not support me; I need to go to work. He reminds me that a job awaits me at Kerman's furniture store. At first, I refuse. Kerman works with a crew of guys in their twenties who seem defeated, guys who have run out of options or who seriously don't care. They go through the motions from nine to five, living to blow off steam after work by sitting around the furniture warehouse, sharing a pint and smoking, occasionally playing cards or craps. I don't drink, I don't smoke, I don't have any money to gamble with, and I absolutely don't want to spend every day of the rest of my life wrestling couches and armchairs on and off a furniture truck.

My father gets me a job at an auto repair shop. I work in the grease pit: changing oil, lubricating car engines, rotating and changing tires. I come home stinking of motor oil and rubber. I quit after a week and grudgingly agree to work at the furniture store with my brother.

Sometime around my eighteenth birthday — the most depressing birthday of my life — I tell Barbara I'll marry her. I don't really propose; I just blurt out that I guess we should get married. She hugs me and seems overjoyed. She sets a date, October 4, and finds a small church and a minister who will marry us. She invites her mother and a few relatives to the ceremony. I invite no one. I can't remember who attends the wedding from my side, other than my mother, who sits in a pew near the back. After the wedding, I formally introduce her to Barbara, her new daughter-in-law. My mother smiles politely and says something pleasant, but in her eyes I see nothing but pain. A few days earlier, she said that we both should try to give Barbara the benefit of the doubt. The word I heard loudest was "doubt."

• • •

I move in with Barbara and her mother. I don't want to share a bedroom with Barbara, but I have no choice. Their place is cramped and claustrophobic, with a tiny living room, a mini-kitchen, her mother's small bedroom, Barbara's room, and one bathroom for the three of us. Barbara, hugely pregnant and virtually immobile, plants herself on the living room couch and spends her days snacking and flipping through magazines, except when she's in the bathroom, which is often. This enrages Barbara's mother, who wails about her lack of privacy, claiming she needs time to prepare herself for work.

Pretty much everything enrages Barbara's mother. From the moment she wakes up until she heads out the door for work, and then again from the second she walks in after work until she stumbles into bed, she rails about her miserable life. She never speaks at a normal volume. She bellows a litany of earsplitting complaints, many directed at me. I don't bring in enough money. If I don't start bringing in more money, and soon, they will turn off — choose one — the heat, gas, electricity, water, phone, or maybe all of them. I'm clumsy, underfoot, always in the way. I eat too much and talk too little.

I begin most mornings with a killer headache. I arrive at work with my head pounding, finding no comfort, no escape there. Loading furniture onto trucks at the warehouse and then delivering and unloading the furniture doesn't just bore me; it deadens me. The work burns a hole in my soul.

Kerman and I, working side by side, don't really say much to each other. Victims of similar circumstances, we live in our own private hells. To get through the day, Kerman takes secret swigs from his bottle and loses himself in country music, which I find an acquired taste. Kerman keeps the radio blasting all day. The twang of the country-western gui-

tar moves him, and the lyrics speak to him. He sings along loudly to
songs like "Hey, Good Lookin'," by Hank Williams, "The Wild Side
of Life," by Hank Thompson, and "A Full Time Job," by Eddy Arnold.
I love music, but I find this stuff grating, only adding to my bruising
headaches.

As much as I hate working at the furniture warehouse, I dread going back to Barbara, her mother, and that suffocating little house even
more. The verses of "A Full Time Job" trapped in my head as I make the
long climb up the hill to Barbara's front door, I sneak inside the house,
bracing for the sound of Barbara's mother's shrill voice, pretending that
I can make myself invisible. My legs heavy, my bones aching, my heart
broken, I want only to collapse somewhere and die.

One evening a few weeks into our marriage, as my mother-in-law
crashes around the kitchenette, shrieking about this or that, I leave Barbara asleep on the couch and slip out the front door. I edge down the
hill, find a spot on the grass, sit, and stare at Spingarn High School, below. Students stream into the gym. A bus pulls into the parking lot: the
opposing team. Kids carrying gym bags and wearing letter jackets get
off. I hear them laugh. I feel their anticipation as they file into the gym.

Then the Spingarn team jogs into the gym. Even from here I recognize everyone. They are guys I know, guys I played with or against, guys
who should be my teammates now. I hear a crowd cheer, a band play,
and the game begins. After a few seconds, the crowd roars. Someone has
scored the game's first basket.

I sit transfixed on that hill, staring down at that school. Spingarn
spreads out right below me, and yet completely out of reach, so far away
it might as well be in a different world.

I have no idea who I am anymore. I'm nobody. A stranger in my
own life.

The late October night turns cold. I shiver.

I've been gone too long. I need to return to my . . . my pregnant
wife . . . and my mother-in-law . . . I have to be up early for work . . .

But I can't move. I shiver again and my bottom lip trembles. I duck my head and stare at the dirt. Below me the crowd roars again, signaling another score, someone making another basket.

RABBIT . . . *Rabbit* . . . rabbit . . .

The words recede, fade out.

I start to cry. Softly at first, and then I lose it. I wail, uncontrollable sobs racking my chest.

•   •   •

Every night Spingarn plays at home, I sneak outside, sit on my spot on the hill, and watch the basketball team jog into the school. And every night my head drops onto my chest and I cry. I do this for two full weeks. Then, one night, I don't go to my spot on the hill. That night I go home. That night I walk into my house on Heckman Street, pause in the doorway, and see my mother sitting at the kitchen table. She springs up from the table and throws her arms around me. I grip her with all my might, never wanting to let go, and then I start to sob.

"I can't do it," I say, my voice halting through my sobs. "I can't stay there. It's too much."

"I know," my mother says. And then she adds, with a force I don't expect, "We're going to make it okay."

•   •   •

That night, I tell my mother everything, every detail: my mother-in-law's ferocious nonstop shouting and complaining; the cramped quarters in that tiny, miserable house, where for the past month I have gotten exactly no sleep; Barbara's total shutting down and shutting me out, her silence a stunning counterpoint to her mother's bluster; my mentally crippling, headache-inducing, physically exhausting dead-end job; inhaling Kerman's secondhand smoke and enduring his painful country music. All of it.

My mother listens and nods, as if she has been expecting me to show up tonight. I see her mind working, calculating, planning. I pray that she's devising an escape route for me, that she will figure out a way to save me from Barbara and her mother. We stay up past midnight, talking and crying, both of us. At one point, my father comes in from his night job driving a cab. He looks at me, raises an eyebrow, but says nothing. He catches my mother's eye and the two of them share a look of resignation. Later, I will relive this moment and decide this is the closest my father ever came to admitting to my mother, "You were right. He never should have married her." But on this night he wearily goes off to bed, leaving my mother to work everything out, to once again protect me.

I move back home the next day.

I call in sick to work in the morning. In the afternoon, I go to Barbara's to talk to her and collect my stuff. I tell her that I will continue working and I promise to support our child. I will honor my responsibility. I tell her that my family will take care of her, but, being private people, we want to keep the birth of our baby as quiet as possible. Because her mother works full-time, and I will no longer be living there, Barbara needs someone to look after her. My good-hearted sister Gladys has offered to have Barbara move in with her and Simiel in Harpers Ferry and stay with them until she gives birth. Barbara will have the best possible care.

"Wait a minute," Barbara says. "You won't be living with us?"

I tell her that not only am I moving out, but I also want the marriage annulled.

Barbara wants to know what that means.

Voided, I say. Erased. Like it never happened.

I tell her that since the day we exchanged our vows at the church, our relationship has not even resembled a marriage. We've been living together almost a month, and we certainly haven't consummated the marriage. Not even close. Not with her mother always lurking and screaming at the top of her lungs. Not that I would ever want to consummate the marriage.

Barbara looks at me blankly.

We haven't slept together, I explain. We haven't *done* anything.

"We don't have a marriage," I say.

She turns away from me and stares at the wall. "We got married," she says. "I'm not giving you a divorce or anything else. You're going to be a *father*."

I say nothing. I finish packing and take a long last look at Barbara. I shake my head sadly. "I'm going home," I say.

At the end of October, Barbara moves in with Gladys.

I continue working at the furniture store, still breaking my back loading and unloading furniture, to the constant twang of Kerman's country music. But living back at home — living apart from Barbara and her mother — I feel as though a block of granite has been lifted off my chest.

On November 9, 1952, less than six weeks after our wedding, Barbara gives birth.

A few days later, when she regains her strength, Barbara leaves my sister's house in Harpers Ferry and, I assume, moves back into that small house above Spingarn High.

I don't see Barbara, and I don't see my daughter. Yes, my daughter. Gladys tells me that Barbara gave birth to a baby girl. I ask about Barbara's and the baby's health, and Gladys reports that mother and child did extremely well. "I had no problem with Barbara," Gladys says. "She was perfectly fine." I don't find out much more — I don't really want to know any details.

I get up the next Monday morning and head to the furniture warehouse to begin a new week of work for a paycheck I'll never see, because it gets sent directly to my de facto wife, the mother of a child I doubt I will ever meet. Each day is an exact replica of the day before and a photocopy of the day to come, my life an endless slow-motion loop.

Thanksgiving arrives. Christmas follows. My family all gets together to bring in the new year, 1953. I get up off my chair and walk into another room, and somebody points out that I've begun to slouch.

"Carrying the weight of the world on my shoulders," I say to no one, to the room.

"Speaking of weight, Elgin," someone says, "you look *gaunt*."

"I don't feel like eating."

"Well," someone says, raising a glass, "to a new year, a better year."

"I'll drink to that," I say, clinking Columbia's glass with my glass of soda. I fight back the onrush of tears I feel building. I don't want to cry here, in front of everybody.

Happy New Year, Rabbit.

· · ·

I have to go back to school.

I know I have to continue paying child support to Barbara. I'll keep my promise. I'll work after school, nights, weekends, whenever I can. But for my mental health, I need to go back to school—and basketball.

My mother picks up the paperwork at Spingarn High, I fill out the forms, and my father signs them. As second semester begins, I transfer to Spingarn. Practically before I sit down in my first class, the basketball coach invites me to join the team. The next day, I go to practice. The players and coach surround me when I walk into the gym. I almost want to drop to my knees and kiss the polished hardwood floor. One of my teammates shouts, "Rabbit!" and tosses me a basketball. I catch it, caress the stitching, and inhale that sweet scent of leather. I palm the ball and hold it out at arm's length. My teammates laugh. My face breaks into a grin. I smile all the way through that practice.

Once we start playing games, I play out of my mind. I'm on a mission. I need to make up for lost time. I average more than 35 points a game. In every game, I feel like I'm elevating to a new height; one time, my hand soars a foot above the rim. I'm so giddy playing basketball again that it feels as though my feet never touch the ground.

I score 31 against Phelps, my former school, and we beat them by 49 points. I sit out most of the second half. *The Washington Post* actually starts covering a few of our games. Early in February, the paper names me Division II Player of the Week and calls me "unstoppable." A week or so later, a *Post* sportswriter comes to our game against a well-known Catholic school. The article he writes is over the top. He gushes about me, says I am "the greatest high school player of all time," and urges, "Don't take anyone's word about him. See him for yourself."

My mother clips the article and insists that everyone in the family read it. Even though I'm embarrassed by the attention, I'm happy that the *Post* has finally recognized our team. Well, to a degree. The next time they list the top ten high school teams in the city, even though we're undefeated, they rank us tenth. At least they've acknowledged us. And then I notice that the nine teams ranked higher than us are all white high schools. The top is Western High. Their star guard, Jim Wexler, set a D.C. record by scoring 52 points in a game. A Washington newspaper calls this the greatest performance in high school history.

Fifty-two points in one game. That's a lot of points. I file that number away.

I'd like to play against that guy. I wonder how I'd do against him.

And then I think: *I wonder how he'd do against me.*

• • •

The basketball season ends in March. We come in second in our league, losing a tough, close championship game. Not bad for Spingarn's first year. Mother and Columbia come to as many games as they can. Pops never makes one. I guess he still considers basketball "foolishness." He never says that directly to me, but he doesn't have to. As Mother always says, actions speak louder than words.

We get a week off for spring break, and I take a few days to visit Gladys and Simiel in Harpers Ferry. After dinner one night, as I'm cleaning up the dishes, Simiel, who's always very quiet, his forehead constantly furrowed as if he's trying to solve a very complex math problem in his head, asks if he can speak with me. After I dry off the last dish and put it away, I go into the living room and find Simiel sitting forward on the couch. I take the chair across from him. He seems to be organizing his thoughts, then he says, "Have you seen Barbara?"

"No. Don't want to."

"I can understand that." He pauses again. "Seemed like she was in a real hurry to get married."

"She was."

"Have you seen the baby?"

I lower my head. "No. I feel guilty about that. I send as much money as I can."

"Elgin, when did you *meet* Barbara?"

"At least a year ago. Let me see—"

"I mean"—Simiel inches to the edge of the couch—"when did you sleep with her?"

I blink. "When did I sleep with her?"

"When did you have sex with her? You need to tell me as close to the exact date as you can. Try to remember."

"It only happened one time," I say.

"When was it?"

Of course, I know. I'll never forget that day.

"April Fools' Day," I say.

"April 1," Simiel says. He smiles at me as if I've solved that complex math problem.

"Why?"

"Barbara gave birth November 9." He then speaks very precisely, emphasizing every word. "To a full-term baby. Not premature. *Full-term.*"

"Okay . . ."

"Elgin, counting from the day you had sex with her to the day she gave birth? That's only seven months."

I know what he's about to say next, but I don't dare speak.

"This can't be your baby."

"Are you . . ." I stammer. "Are you sure?"

"*Yes.*"

The word hits me like a slap.

"I had a feeling all along," Simiel says, almost to himself.

"How do you know she gave birth to a full-term infant? Were you there? Did you see the baby?"

"No. I wasn't at the birth." Simiel slides back into the couch. "But I know the doctor who delivered the baby and the nurse who assisted. The nurse came to see me a few days ago, because she read the article about you in the *Post*. She was very excited. She said, 'Elgin Baylor. He's an unbelievable player. He's a legend. Isn't he your relative?' I said, 'He's my brother-in-law.'"

My head starts to spin. I exhale slowly to calm myself.

"She knew about Barbara because I brought her into the hospital," Simiel says. "She asked me how the baby was doing. I said, 'As far as I know, fine.' And then I said, point-blank, 'I would've heard if there were any complications . . . since the baby was premature.'"

Now Simiel swallows. "She said, 'The baby wasn't premature. The baby was full-term. Healthy and fairly large.'"

Simiel waits a beat and then delivers what sounds like a prosecutor's closing argument. "This baby cannot have been full-term and be your baby."

I bite my lip and hear myself say, "I'm not the father."

"No. You are not the father."

A few months ago, I would've cried, but I feel cried out, stopped up. Simiel, my lifesaver, places his hand on my arm. "Barbara lied to you. She got you to marry her under false pretenses." He slides forward again on the couch. "We need to get a lawyer."

On May 13, I file a complaint for annulment in the U.S. District Court for D.C. I state in my complaint that Barbara gave birth to a full-term infant on November 9, 1952, lied about my being the father, and tricked me into marrying her.

A month later, Barbara files a complaint against *me*. She swears that I am absolutely the father, and that she did not give birth to a full-term infant. She lies about everything again. In an official complaint that she *signs*.

I want to scream. I want to punch my fist through a wall.

Simiel tells me not to worry. He says that both the doctor who delivered the baby and the nurse who assisted in the birth will testify on my behalf.

I feel better. I believe that justice will prevail.

The judge rules in my favor.

Barbara appeals the judge's decision. Twice.

Back in court, after considering our complaints, Barbara's appeals, and all the evidence, the judge issues his final, irrevocable decision. But before he slams down his gavel and rules on my behalf, he singles out Barbara across the courtroom, and says, "You should be ashamed of yourself."

It's over. I win.

The court officially annuls my marriage to Barbara.

I'm a free man.

It's August 1955.

More than two years later.

• • •

I'm on a quest to make up for lost time. Senior year at Spingarn, I average more than 36 points a game and a mess of rebounds. I don't know the exact number — nobody keeps rebound totals in high school — but I bet I grab more than 20 a game. I'm not a stat guy. I just want our team to win. I never focus on the number of points I score or rebounds I grab. Except once.

A year later and I'm still thinking about that number: 52.

The number of points Jim Wexler scored to set the D.C-area record.

I'd kind of like to see if I can break it.

I wouldn't care so much, except that one day a family friend, Curtis Jackson, who works at the post office and always reads the papers, became so fed up with the lack of press coverage our team got that he went down to *The Washington Post,* stormed into the newsroom, and complained to the sports editor. I don't know exactly what Curtis said, but he came back fuming. The person he spoke to had said that the *Post* doesn't have enough space to cover every high school in the city. We all knew what that meant. They don't have enough space to cover the *black* high schools.

I know exactly how much space the *Post* gave Wexler when he set the record: two banner headlines and a full article. They seemed to find plenty of space to write about the white kid who scored 52 points in a high school game. The writer called it a great achievement.

Well, if it's such a great achievement, maybe I should try to beat it against my old high school, Phelps.

Nothing personal against Phelps. They just happen to be the next team on our schedule.

I go after the record early. I score 31 in the first half, but the refs saddle me with four fouls. It almost feels as if they don't want me to break the record. I don't let those four fouls stop me. Coach and I both want to see how long I can play and how many points I can score before the refs hit me with my fifth foul and I foul out.

I never do get that fifth foul. But I do keep scoring. And once the game is well in hand, instead of taking me out, as Coach usually does, he leaves me in. At first I have no idea why. I've been focusing so much attention and energy on playing the game that I nearly forgot about Wexler's record. But then I start to wonder if I have the record in reach. During a timeout, I ask Coach.

"You don't have it in reach," he says. "You already broke it. You got 53. Go for 60 and I'll sit you."

I get to 63 and Coach takes me out, to a standing ovation.

Sixty-three points. I've beaten Wexler's great achievement. The following week, the *Post* does write about it: they find a tiny spot in the corner of the sports section and cram in a couple of lines saying that I broke Wexler's record.

That's how it goes in 1954 in Washington, D.C. I have nothing against Jim Wexler. I'm sure he's great. But like I said, I wouldn't mind playing him.

I end my high school career at Spingarn. We do well at a city tournament for black high schools, finish in second place. I win a trophy for best player in the tournament, get named first team All-Met, and win the Livingstone Trophy as the area's best basketball player. The family gushes over me. Even Pops smiles a little when I come home with that trophy. He's still never seen me play, but he no longer refers to my basketball playing as "foolishness." Mother has turned him around by beginning a new crusade: helping me get into college. She insists that although my marks didn't break any records, basketball may be my ticket. Several recruiters and coaches have already inquired. I've even gotten a call from the Harlem Globetrotters.

Then, on March 12, 1954, while Mother and I are waiting to hear if I'll be admitted to any colleges, I make Washington, D.C., history.

Sam Lacy, the sports editor of *The Baltimore Afro-American,* arranges for a showdown, a matchup between a team of D.C.-area white high school all-stars and my club team, the Stonewalls. It will mark the first time a black team and a white team have ever played each other around here. Lacy asks Bill McCaffrey, a top high school senior, to put together the white team, which Lacy names the Scholastic All-Stars. Of course, McCaffrey asks Jim Wexler, who'd graduated high school, if he'll play. According to what I hear, Wexler says, "I'll bring my sneakers. Just tell me where."

McCaffrey has no problem finding players. Apparently, all the best white players in the city want to go up against me. They're itching to see if I'm as good as my reputation. Then McCaffrey hits a snag. When the principals of these players' schools find out that their white students are scheduled to play a team of black players, they threaten to expel them. McCaffrey doesn't want to put these kids in jeopardy, so he recruits a team of white players who've all graduated high school. Except for one: McCaffrey. He alone refuses to allow the principal of his school to intimidate him. He ignores the threat of expulsion and plays in the game.

Sam Lacy and the *Afro-American* make all the arrangements. They set the time and date, hire refs, and hold the game at Terrell Junior High School, charging a small entrance fee. They don't advertise the game —they don't have to. Word spreads. More than fifteen hundred people show up, almost all of them black. The gym holds only nine hundred, so hundreds get turned away. Later I'll hear that several people sneaked in. Some climb in through windows. When we arrive to play what the paper calls a "mixed basketball battle," the gym is a roaring, rocking madhouse.

We take our places on the court to prepare for the center jump that starts the game. Wexler and I greet each other at midcourt, shake hands, and say in unison, "Good luck."

I don't say much after that. I disappear into a kind of zone. Even though there are people everywhere, crushed against walls, overflowing the bleachers, and police officers roaming around, I block out everything except the players on the court, the hoop, and Wexler.

The Stonewalls take charge early. We never trail and we never look back. I hit my first eight shots. I shoot from outside, I drive to the basket, I fake, I float, I fly. Once, with Wexler on my hip, I spin, turn my back on him and the hoop, grab the ball with two hands, leap, and dunk the ball over my head — backward.

For a full second, the entire crowd goes silent.

Then I hear a murmur.

And then I hear a deafening roar, like a bomb going off.

The gym shakes.

I don't think anyone had seen a reverse dunk before. I had never *done* a reverse dunk before.

I run past Wexler on my way back downcourt. I catch his eye. He wags his head and smiles out of respect, out of awe. I don't say anything. But I give him a wink.

Wexler plays well. He scores 34.

I score 44.

The Stonewalls smash the Scholastic All-Stars by 25 points.

An historic game, blacks against whites for the first time. *The Baltimore Afro-American* is the only paper that covers the game.

Years later, Jim Wexler will recall our game and speak in reverence about that reverse dunk. He'll say about me, "He showed me basketball at a totally different level — another world, heads and shoulders above anything I'd ever seen. He could do everything."

4

COYOTE

AUGUST 1954.

Mother sobs. I don't want her to cry.

I close my eyes, cough, and clear my throat. Now I don't want her to see *me* cry.

Mother clasps me in a death grip, her arms around my middle, squeezing me so tight I'm afraid she'll crack a rib. Towering over her, I pat her back gently. I open my eyes. Pockets of people surround us on the platform, some of them crying, too. Other people scurry by, lugging suitcases.

My mother muffles another sob.

"I'm just going to college," I say. "I'll be back for Christmas."

"Idaho is so *far.*"

"I'll be all right." I look over my mother's head and catch my friend Warren Williams, known as W.W., who stands a few feet away, looking down at his shoes, waiting for me to finish saying goodbye to my mother so we can board our train to Chicago, the first leg of our three-day journey to Caldwell, Idaho. "And I got W.W."

At the sound of his name, W.W. lifts his head and waves.

A voice on the loudspeaker announces our train.

"That's us," I say.

My mother nods, pushes off from my chest, dabs both cheeks, and smiles. Instantly, it seems, she's pulled herself together. "You got your money?"

"Yes."

"It's not all in one place? Did you spread it around? Hide some of it in your shoe?"

"Don't worry."

"I do worry. I can't help it: It's my job. I'm your mother."

I nod, my tears starting to come.

"I'm proud of you, Elgin," Mother says, touching my cheek. "And I know you'll be all right."

I nod again. Suddenly I feel as if everything in my life has changed —has flipped—in this one blip of time. I know instantly that I'm not just going to college. I'm going away. I'm leaving D.C.

I want to leave D.C. I want to see what it feels like to live in a world that is not starkly divided into black and white. I know that I will be among only a handful of black students at the College of Idaho. W.W. scouted the school and told me that he saw only a few black faces in a sea of white students when he visited. But he said that the people he met were welcoming. Bordering on *nice*. It was . . . different. At the College of Idaho, black students and white students take the same classes, share the same dining hall, play on the same teams. I've had enough of segregation, isolation, and prejudice.

I'm also leaving home.

I want to leave home. I want to separate myself from my father and his anger at the world, mostly the white world. I want to get away from that. I still feel so much hatred toward him. Maybe by going so far away, I will leave my anger and hate behind.

Then, nodding toward our train and seemingly reading my mind, my mother says, "You have to go."

"I know," I say, and hug her again. This time she stays composed and strong. I duck my head. "Thanks for everything," I say.

Those words sound so inadequate. I need to tell her more, need to tell her how much I appreciate her, how much I'll miss her, but I can't find any more words. At least I spoke the truth: My mother has done everything for me. She has been everything.

"Take care of each other," she says, eyeing W.W. and me.

"We will," I say. I retreat a few steps and pick up my suitcase, and
W.W. and I head toward our train. I turn and wave. Mother waves back.
My lip quivers, but I don't cry. I wait until I'm on the train.

• • •

The summer has been a whirlwind.

W.W., a standout football player at Dunbar High School, a friend I
met playing playground basketball, has, along with my mother, taken it
upon himself to get us both into college. Early on, W.W. and I decide
we want to go to the same school and room together, and we vow to get
out of D.C.

Back in April, after the basketball season, coaches from Seton Hall
and Duquesne inquired about me. I also heard that some college coaches
had watched me play for the Stonewalls. I was named captain for the
season, and we entered local tournaments, playing our games at Turner's Arena in front of standing-room-only crowds of as many as two
thousand people. We roared through the first tournament, winning each
game by 20 points or more as I averaged 39 points a game.

In our second tournament, we made it to the finals against an all-
white college team made up of players from Georgetown, La Salle, and
the University of Maryland, among them scrappy guard Gene Shue,
who would go on to become a five-time NBA All-Star and First Team
All-NBA.

Shue was the real deal. We traded baskets to begin the game, and
then Shue hit a tricky twisting layup and the college all-stars grabbed
the lead.

That was enough for me. I took over, scoring 16 points in the first
quarter, but even more, I dominated the boards. We built a sizable halftime lead and ultimately beat the college all-stars by 22. Shue played
well, scoring 34. I scored 38.

I graduated Spingarn in May. As we headed into June, the Stonewalls
stepped up their summer schedule. We played several semipro teams,
and then we agreed to a rematch against that same team of white college

players. I got word before the game that some guy on their team had been mouthing off, guaranteeing that I wouldn't score 38 points against *him*. He was right: I scored 47. I'm not a big fan of bragging.

Over the summer, I got more inquiries from college coaches, but no scholarship offers. I think my marks scared away some of the schools, and perhaps my dropping out of high school, falling off the radar while I worked at the furniture store, and then transferring to another school confused others. I couldn't explain that my grades had suffered and I'd dropped out of school because I'd been tricked into a marriage, which I was in the process of getting annulled.

As August approached, W.W. and I still hadn't found a college that would take us both. Then W.W.'s father heard about an opportunity for us at tiny College of Idaho, student population 450. Because of the school's limited athletic department budget, Sam Vokes, the football and basketball coach, tried to recruit talented athletes who could play both sports. Two years ago, he enrolled R.C. Owens, an All-American wide receiver and an excellent basketball player. Owens reinvigorated the basketball program to some degree, and he put the football team on the map. It turned out that Sam Vokes was well aware of W.W. and liked what he'd heard about me. He may have read a newspaper clipping or two. He pitched me hard on the school and the sports program, and then offered W.W. and me scholarships to play football and basketball. I'd never played organized football, only sandlot, but Vokes didn't seem to care.

· · ·

The train heads west, then north, departing D.C. at four in the afternoon, scheduled to arrive in Chicago the next morning. W.W. and I can't afford a sleeper car, but we find a couple of empty rows in a regular car and stretch out. As night falls, the steady rumble of the train lulls W.W. to sleep. Wide awake, I press my face against the window and watch our country roll by, a slide show in night shadow—a forest, a small town, a block of factories, a cornfield, a dark plain. I've never been outside of Washington, D.C., other than to visit Gladys and Simiel in Harpers Ferry. Now I'm on my way to what feels like a foreign country. W.W., who's spent some time in Caldwell setting us up, has tried to explain what to expect, but we both agree I need to see it to believe it.

I look down at my dress pants, one of two pairs I brought, the crease sharp as a blade. I ironed my pants and shirts and folded my sports jacket meticulously the night before we left. I've packed my suitcase with almost everything I own—my dress clothes, my basketball gear, my sneakers, my portable record player, and a few records I've slipped between my underwear and socks. Sam Vokes warned me to prepare for a harsh winter. By October, he said, the weather in Idaho can turn bitter, windy, and biting cold. I'm prepared. My brother Sal, planning for a life in California, has bequeathed me his overcoat.

As I peer out the window, I suddenly feel disoriented. Not only do I not know what to expect at college, but I don't even know how to *be*. The past few weeks of my life have accelerated at breakneck speed, and now I don't know where I'm going, literally. I stare at my reflection in the window and I think, *It's so strange. You're going to college to play football.* I look myself in the eye and say aloud, with true uncertainty, "I just hope I can make the basketball team."

Basketball, I realize, is all I know. Basketball is who I am. The basketball court, not some college, is where I feel safe.

The next morning, sometime before noon, we arrive at Chicago Union Station. I've managed to catch maybe two hours of sleep, and I wake up feeling sore and hungry. W.W. and I drag our suitcases off the train and lurch into the main area of the station. We suddenly stop, dwarfed and dazzled by the size and brilliance of what we see: a white marble floor extending to the far end of the horizon, a hundred-foot shimmering glass atrium rising into the sky, the station itself the size of two city blocks.

W.W. taps my arm and points at the men's room.

"I'll watch your bags," I say.

He trundles off and I stay pinned to my spot, bathed in the golden light from the vaulted skylight.

"You got the time?"

A fidgety man wearing a rumpled overcoat appears next to me. He flashes a smile. I see that he's missing two front teeth.

"Don't have a watch," I say.

"That's too bad." He nods at the suitcases at my feet. "Going on a trip?"

"Going to *college*," I say, emphasizing the word with some pride.

"Good for you," the man says. "I expect you will need to know the time in college. You don't want to be late for class. Or an important *date*. You could use a watch."

"Well, maybe I'll get one."

"I happen to have one," the man says. He looks over his shoulder as if he's expecting someone, then whips open his coat. At least two dozen watches dangle inside, pinned to the lining.

"This one here," the man says, removing a watch and placing it across his palm, "is ideal for a college student. It's very handsome on the wrist and keeps perfect time. You can wear it for any occasion—class, dress-up, anything. Cost you only thirty dollars."

"Thirty dollars?"

"Try it on. See how it feels."

In one lightning-fast motion, the man straps the watch onto my wrist and fastens it in place.

"That looks great on you," he says. "Very dashing. Now you'll never be late. And, like I say, it keeps perfect time. Just make sure you wind it every couple hours or so."

"Thirty dollars is a lot of money . . ."

"Not for this watch. This is a very expensive watch. It's a Timex, I believe, or, hold on, it may be an Elgin—"

I almost burst out laughing. "An Elgin, huh?"

"Could be. The brand name seems to have rubbed off . . ." He glances over his shoulder again. "I'll give you a deal because you seem like a nice kid. Twenty-five."

"Here." I reach into my pocket, pull out a crumpled bill, and hand it to him.

The man unfolds it and frowns. "This is a ten."

"Take it or leave it," I say.

"Twenty," the man says.

I unfasten the watch and hand it back to him.

"Fine. Ten." He snatches the bill, buttons his coat, and hustles away.

I strap the watch back onto my wrist and smile.

I know I've been ripped off. I'm sure the watch is probably worth two dollars, maybe less, certainly not ten, and I know it's no Elgin or Timex. But I have to have it. (I actually still have that watch and it still keeps perfect time, as long as I wind it every couple of hours.)

Standing in the magnificence of Chicago Union Station, I know that man has appeared out of nowhere and handed me a symbol:

It's time to start my new life.

. . .

A day and a half later, our train pulls into a quaint station in Caldwell, Idaho, on the edge of the College of Idaho campus. I stand up slowly, stretch my throbbing back, and peer out the window at a tiny, sleepy town.

W.W. and I button our dress shirts, tuck them into our pants, knot our ties, put on our sports jackets, and climb off the train. Lugging our suitcases and squinting into a bright midafternoon sun, we begin a slow walk through the center of campus, toward the apartment building where we will live, across the street from the main quad. Walking through campus, I take in pristine brick buildings and a large swath of freshly mowed grass dotted with lightly swaying, fragrant bushes and manicured trees. The air smells crisp and has a scent I don't know, maybe evergreen. A steady stream of people our age walks toward us, every one of them white, every one dressed in jeans and T-shirts, every one greeting us with wide smiles, waves, or hellos.

Is this place real? Everything here feels so clean and so . . . *white*. Even the park benches scattered around the campus are white. Years later, when I think about the College of Idaho campus, I'll picture the 1950s

TV show *Leave It to Beaver*. It's as if W.W. and I have wandered into a private and exclusive members-only club, except that, rather than feel intimidated or excluded the way I do in D.C., I feel invited.

Two young white men wearing jeans and T-shirts approach, smile, wave, and say hi, all at once. W.W. and I smile and wave back. I consider my creased dress pants, dress shirt, sports jacket, and necktie. I look like an insurance salesman. I breathe in the scent of evergreen wafting all over the place. I shake my head at all the lush greenery. "What *is* this place?"

W.W. yanks off his tie. "I don't know. But we are way overdressed."

"Are we the only black people here?"

"No. There are seven of us." He grins. "When I was here over the summer, I counted."

By the end of the first week, I've settled into a furnished room on a floor that I share with several other black athletes, including one-armed Gary Mays, the Bandit, who has been recruited from D.C. to play basketball and baseball, and Raleigh C. Owens, known as R.C., an upperclassman and football All-American from Louisiana. When R.C. eventually joins the NFL and becomes an All-Pro wide receiver for the San Francisco 49ers, he will earn a new nickname, "Alley Oop," for his ability to leap high over defenders in the end zone to snag touchdown passes. R.C. becomes our tour guide and college and town insider, advising us on which classes to take and which to avoid, showing us short-cuts through campus, where to eat in town, where not to eat because we would not be welcome, and, highly important, where to find the best parties on weekends. W.W.'s count of black students turns out to be off by at least ten, although I see only two black girls on campus. Socially, it doesn't matter. The white kids enjoy partying with black or white kids. Seems the white kids enjoy partying, period. That's about all there is to do in Caldwell, Idaho: party, play sports, and study. I throw myself into all three and do all right. I pull a B average in my courses and get straight A's in sports and partying.

It rains the first day of football practice, and Coach Vokes herds the team into the gym. We stretch and do a few exercises, then he divides the team according to position. Assuming I'm a wide receiver, I follow R.C. into a line of tall, gangly guys and fall into a passing drill. I don't know what I'm doing, so I copy R.C., cutting this way, slanting that, catching passes from our starting quarterback. After a while, I not only get the hang of it, but I get *into* it. I start imagining myself as a wide receiver lined up on the opposite side from R.C. He goes about six foot four and, with me at six-five, I see us as an unstoppable double threat. I'm sure I could play football. I can run, I've got good hands, good moves, I'm pretty strong. How tough can it be?

Coach Vokes blows his whistle to end practice and starts stuffing errant shoulder pads into a fat duffel bag as the team begins filing out of the gym. I'm about to head out myself when R.C. shouts my name and tosses me a basketball he's found tucked into a corner. We start shooting around, and then a few more guys join us and we make teams for a pickup game.

We run full-court. I take my time at first, sort of lope, and then I pick up my pace. I streak toward the hoop, slide along the baseline between two defenders who converge on me, pump, fake, fly past them, and bank in a reverse layup.

"*What?*" R.C. says from across the court. "I never saw that before."

The other team runs it back. One of their players tries an outside shot. I jump, block the shot, tap the ball to myself, and dribble toward our basket. Gary Mays, shirtless, sweaty, decides to check me, close. He bodies me up. "Don't touch me with your nub. You know I hate that!" I say, then motor around him, stop at the top of the key, and drill in a jumper.

A couple of guys react. And then, in my peripheral vision, I see that Coach Vokes has stopped stuffing shoulder pads into the duffel bag and is standing, hands on hips, intently watching the game.

Our team goes up by three baskets and then R.C. switches over to guard me. The other players on the court clear out, isolating the two of us. For this one moment, all the other players become spectators. R.C. hand-checks me, presses my hip, and shoots me a half-smile, half-taunt. "Come on, *Rabbit*, let's see what you got."

R.C., now a junior, enjoyed a memorable sophomore season on the basketball team. He averaged 27 points a game and led the conference in rebounding. Now he's challenging me, a freshman, getting right up in my face, directly in front of the basketball coach.

I don't talk trash, but I do smile thinly and say, "I got a little something for you."

I don't plan anything special. I don't plan, period. I just go. I eye R.C., give him a little shoulder shimmy, then fake right, fake left, and then *go* right. R.C. commits on the first fake and almost trips over himself trying to stay with me. I take off from just inside the foul line, soar, and slam the ball through the hoop.

Everyone screams, including R.C., still standing back near the foul line. He shakes his head and offers me his palm, which I slap. "I am looking forward to playing with *you,*" he says.

Then Coach Vokes jogs toward me, his face a concerning shade of red. He stops and puts his hand on my shoulder. "About the football team," he says. "You're no longer on it."

I look over at R.C. "Coach, the two of us—"

"Let me say this a different way. I can't risk you getting hurt. I'm not *letting* you play football."

"But all my friends are on the team, everybody I know here—"

Coach nods, finally understanding my hesitation. "Don't worry. You'll be *part* of the team. You're just not going to be *on* the team."

"I don't understand—"

"Put the rest of the shoulder pads in that duffel bag, grab any extra helmets, and bring everything to the cage in the locker room." He presses my shoulder warmly. "Congratulations. You're my new equipment manager."

And so I spend the football season checking the team in, handing out and collecting helmets, shoulder pads, kneepads, hip pads, and towels, making sure we have enough water bottles, passing out schedules and other paperwork, and prowling the sideline next to Coach Vokes. I actually don't mind being equipment manager. I like being part of a team. I enjoy being close to the action, and, while I love watching R.C. take

the art of receiving to a new level—he once makes 15 catches in a game
—I'm relieved not to have my body repeatedly pounded into the turf
by adrenaline-fueled three-hundred-pound farm boys.

I prep for basketball season by playing pickup games in the gym.
R.C. joins us most games, and he and I quickly develop a rapport on the
court, a chemistry that deepens when the season starts. As the evening
of our first game approaches, a home game to be held in our Cracker
Jack box of a gym, which holds maybe five hundred people, I feel a buzz
on campus. Last year, the basketball team played in front of depressingly
small crowds, sometimes fewer than fifty people. This year, word is out
— *Come see your College of Idaho Coyotes. We got this freshman from Washington, D.C. This guy can really play.*

We open the season in front of a sold-out gym. I jump center and tip
the ball to R.C. He flips the ball back to me, and then it begins—a kind
of magic. I keep the whole team involved, but when it comes to R.C., I
don't have to look to see him. I just know where he is. We run two-man
fast breaks, streaking ahead of everybody. I hit him with an array of no-
look passes. He has such good hands that he catches every pass I throw,
even when he doesn't know they're coming. He then scores himself or
passes back to me and I score. On the boards, we're a force. I'm a strong
rebounder, but R.C. is a beast. He snatches balls off the backboard like
he's mad at them and wants to punish them. Then, *whoosh,* we're gone
—showtime, before anybody heard of Showtime.

I score 57 points in that game and 46 in the next one. Our team rips
off eighteen wins in a row, finishing with a record of 25–4. I complete
the conference season by scoring 53 in our last game. We enter the division playoffs at a site far from Caldwell, where no one has seen our team
or my style of play.

In the first game, I start strong, with a couple of quick baskets and
a behind-the-back bounce pass to R.C. on a blistering fast break. Then
one of our players steps to the foul line to shoot free throws. He makes
the first and misses the second on purpose. From my hours and hours
spent studying the angles of balls bouncing off too-tight rims and wobbly backboards on the playgrounds in D.C., I know exactly where the

ball is going to go. As it caroms off the rim, I dart in front of the defender closest to the basket, reach over my head for the ball, and dunk it
in one motion. The crowd goes crazy, but the refs seem confused. They
quickly confer and call me for taking too many steps. A few plays later,
they again call me for traveling. Over the rest of the half, they call me
for steps about half a dozen times. They also display a strong taste for
home cooking, hitting me with four quick fouls. Severely hampered by
fouls, I can't play my game, and we lose. Still, we've had a breakthrough
season, making it this far and pulling in huge crowds. R.C. leads the
conference in rebounds. I finish second. I average 31.3 points a game
and get named a small-college All-American. Not bad for a freshman
playing varsity. I'll take it. And I answer my own question.

I guess I *can* play against this competition.

· · ·

We party, blacks and whites together, Friday nights usually, Saturday
nights always. Most kids loosen up after a drink or two. The white kids
love beer. I'm stunned by how much beer they can drink, even the girls.
None of the black kids drink beer. Personally, I can't stand the taste.
Most black people I know prefer hard liquor or something sweet, like a
mixed drink or spiked punch.

We put on records and turn up the volume, and then the party rips.
I select records from my stash, blasting my favorite, Big Joe Turner,
the "Boss of the Blues," and dance the jitterbug to "Shake, Rattle and
Roll," singing along with Big Joe, "Get outta that bed, wash your face
and hands . . ."

One day, as a bunch of us meander into a classroom and take our
seats, a girl I've seen around, a white girl, sits down next to me. She
smiles and asks if I'm going to the party Saturday night.

"Definitely," I say.

"Great. Me, too." She blushes and lowers her voice. "I've seen you
play basketball, and I've seen you dance. You're really good."

"Oh, yeah? Well, thank you."

"Where did you learn?"

"On the playgrounds, and I played in high school—"

"No, where did you learn to *dance?*"

"Oh. I don't know. At parties. My sister taught me some of the dances. The jitterbug. Used to call that the Lindy hop."

"Would you teach me?"

Now, this is new.

Where I grew up, in segregated D.C., white girls would never *talk* to a black guy, never mind ask him to teach her to dance. I feel my mouth drop open in shock. It takes me a while to regain the ability to speak, and then I say, "Okay, sure."

"My name's Sandy," she says.

"Elgin."

She blushes. "I know." Then she adds, "I want to learn rhythm."

"Huh. Not sure I can teach you that. But I can teach you the dance steps."

Saturday night, as Big Joe Turner starts growling "Shake, Rattle and Roll," I dance over to Sandy, who stands, swaying to the music, next to two of her friends. I offer my hand and beckon her to join me on the dance floor. She reaches toward me tentatively. Her hand feels cool. I lead her into an open area where someone has rolled up the rug and a few kids are dancing. Badly. They look like they're trying to stamp out a campfire. Holding Sandy's hand, I go into the jitterbug. I demonstrate the basic steps, calling them out: "Side, side, toe tap, toe tap, rock back, rock back, triple step, rock back, repeat . . ." Eyes on my feet, she copies me, hesitates, and then picks up the steps quickly. Once she seems comfortable, I add more elaborate moves. I twirl her once, twice. She follows easily. She feels the music. Don't need to teach her any rhythm; she's got plenty of it. I ease her hand behind my back, spin myself around, twirl her.

"You're a natural," I shout over the music.

She beams. "I've never met a black person before."

"How am I doing?"

Her face pulses deep red. I laugh and swing her around again. The song stops, and we both raise our hands and wiggle our fingers.

Another song starts, "Boogie Woogie Country Girl," and Sandy says something I can't hear. I bend over and she shouts into my ear, "Will you teach my friends?"

By the end of the night, I've taught Sandy, her two friends, and a few other girls the jitterbug and its variations. I've also put on a Perez Prado record and shown them the hot new dance, the cha-cha-cha. I'm not sure why I'm surprised, but the girls learn the dances easily.

The guys?

Terrible. They're clumsy, wooden, and tone deaf. Some of them are so bad I'm afraid they'll injure themselves or maim their dance partners. One guy, a lineman on the football team, slams his boot down so hard trying a dance step that he practically crashes through the floor.

I don't expect what happens next.

The *guys*—the lineman, several friends on the football team, and three basketball players—come to my room and ask me for dance lessons. I tell them all the same thing.

"I've seen you guys dance. I don't think I can help."

"You can teach us. We're coordinated. We're athletes."

I try.

I manage to teach most of the guys a modified jitterbug. I give up on the lineman and offer, as a last resort, to teach him to slow-dance. We clasp hands. He plunks one of his mitts on my shoulder. I rest a hand on his waist to steer him and, miraculously, he picks up a rudimentary box step to a Jerry Butler ballad. Forever grateful, the guys beg me to keep our dance lessons a secret.

"Are you kidding?" I say. "I'm not telling anybody."

One night, two football players, both black, arrive at my door. "Let us in, quick," the first guy says.

They hustle inside and I close the door behind them.

"Look, we're black," the second guy says.

"I see that."

"We're supposed to be able to dance."

"Except we can't," the first guy says.

"Can you teach us?" the second guy asks.

"You need rhythm," I say. "You have some rhythm, right?"

"Sure," the first guy says. "Well, you know—a little."

"He's lying," the second guy says. "We have none." He's right. They have no rhythm at all. I try to teach them the jitterbug, then a slow dance, but they're actually worse than the lineman. "What are we going to do?" the first one says.

"Fake an injury," I say. I look at the first guy. "You pulled a hamstring." I turn to the second guy. "You broke a toe."

They look at each other.

"You're good," the first one says.

• • •

In March, after we get knocked out of the playoffs, I follow the NCAA tournament in the newspapers, and I invite a bunch of friends to my room to listen to the championship game on the radio. The guys sit on the bed, on chairs we drag in from other rooms, and sprawl on the floor as we listen to the University of San Francisco, led by their All-American center, Bill Russell, taking on La Salle University and their star player, high-scoring Tom Gola. I sit, mesmerized by the broadcast, lost in the action, imagining each play. To me, Bill Russell is a phenomenon, a superbly athletic player who dominates each game with his *defense*—blocking shots, ravaging the boards, snapping off outlet passes to teammates in mid-sprint halfway down the court. He averages about 20 points a game, but he seems to score most often when his team needs a basket—or, frankly, when he wants to. San Francisco wins easily, 77–63, on its way to winning sixty games in a row.

As the game ends and the broadcasters announce that Russell has won the award for the tournament's Most Outstanding Player, I wonder:

How would I do against the best college players in the country?
How would I do against Bill Russell?

. . .

Spring break. I fear for my life. I'm in the air. Flying. My first time in an airplane, a Piper Cub, a plane that seems as flimsy as a toy.

We pass through clouds. We dip, we rise, we *shake*.

I hold my breath. We are flying from Boise to Seattle. I ride shotgun. The Piper Cub's engine thunders. The plane rattles. I look through the windshield, completely immobilized, my entire body quivering. Next to me, the grinning, helpful pilot points out the sights below —that remarkable mountain range with snowcapped peaks, that stunningly dense green forest with the tips of evergreens stabbing up at us, those miraculous jagged cliffs, that sky-blue body of water, as placid as glass.

I see none of this. I see nothing. I stare straight ahead, my eyelids fluttering, my heart thumping, my hands gripping my shaking knees, my stomach churning.

This, I realize, is what fear feels like. True fear. I thought I had experienced fear before. I had not —not even close. This feeling? I'm afraid I will die. That's this fear.

"Isn't this breathtaking?" the pilot says. "Oh, look, to your left, that roaring waterfall . . ."

I stare straight ahead. I try to shut out his voice. My mind drifts.

Two weeks ago.

Blindsided. That was what happened. Like being at the beach, facing away from the ocean, and without warning a tsunami roared up and washed me out.

Coach Vokes called me into his office. He told me to shut the door and take a seat. When someone tells you to do that, it's never good. It's usually a prelude to getting fired. But I'm only a student. I can't get fired.

"We're consolidating the sports program," Coach Vokes said.

I didn't know what that meant, but it didn't sound good.

"What?"

"We're concentrating on football."

He took an interminable pause. I spoke, because I couldn't stand the silence.

"You didn't mention basketball," I said. "We had a great year. Got knocked out of the playoffs because of some bad home cooking. Next year—"

"Not gonna be a next year," Coach Vokes said. "We're not gonna have a *team*," he said. "It's a financial thing, Elgin. The university needs to cut back the athletic program." He then added, solemnly, "You almost saved us."

"What do I do?"

He shrugged. "I guess you could play intramurals . . ."

Coach Vokes continued talking, but I didn't hear what he said. The truth is that, while I was shocked by this news, I'd been living in a state of uncertainty since the season ended.

While walking back to our apartment shortly after that, W.W. told me that he and the Bandit were leaving school. He said they were going back to D.C. They'd enjoyed the people in Idaho, appreciated the clean air and stunning scenery, but going to a school of 450 people, more than 60 percent of them girls and only twenty-five of them black, has been, at various times, challenging, uncomfortable, weird, and lonely.

"I miss D.C.," W.W. said. "Or maybe I'm just homesick."

"Is this a done deal?"

"Done," W.W. said.

So, within a few days, I lost my two closest friends and I lost basketball.

For the next twenty-four hours, I walked through campus in a kind of trance. Then rumors started buzzing, rumors about *me*—I was transferring to a major basketball program on the East Coast, possibly La Salle or NYU; I was heading west, going to play at the University of

Portland or UCLA; I was leaving college to barnstorm the country as the newest member of the Harlem Globetrotters.

And then, amid all these rumors, I got an actual call and a serious inquiry, initiated by, or at least encouraged by, Coach Vokes.

Al Brightman, coach of Seattle University, had heard about me. He must have heard all the same rumors, because, to woo me, he got the brother-in-law of Ralph Malone — a prominent Seattle businessman, owner of Westside Ford dealership and a major university booster — to fly me to Seattle in his private plane.

The Piper Cub I now sit in, sweat suddenly cascading down my forehead.

The pilot pointing out all the sights below being Mr. Malone's brother-in-law. "You're going to love Seattle," he says. "Great city. A little more cosmopolitan than Caldwell."

"Urm," I say, eyes straight ahead, fixated on the blank blue sky.

Finally, we land. Still staring ahead, I wait for the pilot to taxi down the runway and pull into his assigned slot before I allow myself to breathe normally. I climb down off the plane. As my feet touch the ground, I try not to scream *"Thank you"* to the heavens.

For the next two days, W.W. and I, guests of Ralph Malone, receive the grand tour of Seattle and Seattle University. I find the Emerald City, as some call it, physically striking, a grid of surprisingly steep hills bordered by Puget Sound, mountain ranges, and evergreen forests. The Seattle University campus sits in the middle of the city, with restaurants, shops, record stores, and student hangouts within walking distance. The city and the university seem alive. More than that. Exciting.

On the second day, I meet Coach Brightman and a few of the players on the team. I've done my homework and I know that they had a very successful year, going 21–4 and winning their league. The team features two or three black players, but their stars are Eddie and Johnny O'Brien, five-foot-nine white twins who, according to what I read, can dunk. I'm not sure I believe that.

"It's true," one of the black players tells me.

"Really?" I say. "I don't think so."

"Come to Seattle," another player says. "See for yourself."

It's so far, I think as I'm flying back to Idaho, allowing sleep to overtake me on the return trip. Completely across the country. Feels like a different planet.

But, as Coach Vokes pointed out, in Idaho, the only people who've heard of me are the people in Idaho. In Seattle, he said, everyone in the country would know me.

• • •

I decide to transfer.

I know it's the right decision, but it's not an easy one. For one thing, I don't want to go to Seattle alone. I don't think I can bear that.

I talk to two of my former Spingarn teammates, Lloyd Murphy and probably my best friend, Francis Saunders, about going to Seattle with me. When they agree, I call Coach Brightman and convince him to offer them scholarships. I tell him that the three of us make up a package deal. All of us or none of us. I don't say it that harshly, but he knows that's the bottom line. He accepts the package.

The other complication is that, according to NCAA rules, because I will be transferring, I will be ineligible to play collegiate basketball for one year. I will have to establish residency and enroll as a sophomore at Seattle University for the 1956–57 season.

Coach Brightman explains this to me before I leave Seattle.

"But you will play basketball," Ralph Malone assures me. "I'll take care of that."

"How?"

"We got an AAU league. You'll play for my team, Westside Ford." He shakes his head. "Tough league. We finished in the middle of the pack last year." He allows himself a tiny smile. "But I'm feeling better about this year."

. . .

Summer 1955.

I return to D.C. to visit my family and tie up loose ends before going to Seattle.

I spend as much time as I can with my mother, and I visit Gladys and Simiel often in Harpers Ferry. Kerman mostly keeps to himself, and Sal has moved to California. At the end of the summer, he will visit and drive me, Lloyd, and Francis to Oakland in his Oldsmobile, and the three of us will catch a train to Seattle. Columbia remains elusive, still living a life in the shadows. I don't see her much. I worry about her a lot. I miss her constantly.

I try to avoid Pops. I arrange to be home when he's working and stay out when he's home. My hatred for him still simmers within me, but now it has taken the form of a dull, gnawing ache. I don't want to see him because I don't know how I will react. I don't want to confront him, and at this point I know he won't say anything directly to me. We will avoid a scene. We Baylors, especially the men, keep our feelings bottled up. We don't express ourselves. We close ourselves off. I, at least, have an outlet: basketball. Playing hours of pickup ball, I take my fury out on the rim or my opponent.

When I do see my father, I'm struck by how he seems smaller, less imposing, not as intimidating. Maybe it's because I've grown physically. I've bulked up, added twenty pounds of muscle from the drills Coach Vokes put us through and by taking a boxing class for my physical education requirement at school.

In August, I appear in court for the last time, officially ending my marriage to Barbara after two years of rulings, filings, court procedures, and appeals. I'm now free from her in every way, including, thankfully, financially. Then, before I know it, Labor Day arrives and, soon after, my twenty-first birthday.

September 16, 1955.

I can vote.

I can drink.

I am legal. I am a man.

The evening of my birthday, I walk into the living room of our house. My father sits in his chair. I stand facing him. He glances up from his newspaper. I say nothing. I turn and slowly walk to the cabinet in the corner where he keeps his cherished bottle of Colonel Lee, the bourbon everyone in the family knows is meant for him and only him.

I open the cabinet and remove the bottle of Colonel Lee and a shot glass. I open the bottle and pour myself a shot.

I feel my father's eyes boring into me.

I take the armchair across from my father and down the shot in one gulp. The bourbon burns my throat. My eyes water. I blink away the tears.

I glare at my father now, that simmering rage rising. I want him to say something to me. I want him to chastise me. I want him to tell me not to touch his precious liquor. *Say something to me, Pops. I dare you.*

He can't look at me. He turns away. I pour myself a second shot. I lean back in the chair and nurse that shot, take my time.

Say something. Come on, damn it. Say something.

He turns back to me. He studies me for a long count of ten. Then he says in a murmur, his voice cracking, "Looks like there's a new man in the house now."

He dips his head and goes back to reading the paper.

I can't hate him anymore. Whatever rage I've held inside me crumbles, pulverized into a kind of emotional dust. I feel sad for my father now. And, strangely, I feel close to him, closer than I have ever felt. At this moment, I feel as if I understand how this man has been formed by degrees of fear, confusion, pride, and outrage, as well as by a deep-rooted instinct to survive. I even understand the impotence he must feel when it comes to fighting against forces larger than him, larger than all of us. He's right. I have gone away, and I have grown up. I have become a man.

I return the bottle to the cabinet and walk out of the house.

5

CHIEFTAIN

SEATTLE. FALL 1955.

Johnny O'Brien, player-coach for Westside Ford, doesn't want me.

"We don't need another player," he says on the phone to Ralph Malone. "We already have eleven guys."

"Do me a favor," Mr. Malone says. "Let him practice with you. See what he's got."

A few days later, I walk into the gym where the Westside Ford team holds practice, pick up a basketball, and shake hands with Johnny, a former Seattle University great who, at five-nine, played center.

"I read all about you and your brother," I say.

"Don't believe everything you read."

I wink at him. "You're not identical twins, right? He's taller."

He practically shouts. "*I'm* taller. By a lot. Quarter of an inch."

"My mistake." I smile. "I hear you can dunk."

"That actually is true."

I nod, rotate the ball in my hands to get the feel of the stitching, dribble a couple of times, turn, sprint toward the basket, and throw down a two-handed dunk. I retrieve the ball, head back to Johnny, and hand him the ball. "Now you."

Johnny cracks up. "Maybe later," he says. "Much later."

I will eventually see Johnny O'Brien dunk the basketball after a running start, but at that moment he can't stop smiling. "Our first game's tomorrow at eight. You can make that, right?"

"It might be tough. I don't have a car."

"I'll pick you up."

For the next nine months, I play alongside Johnny on Westside Ford, in the Northwest AAU League, against teams like the Buchan Bakers, Puhich Cleaners, and Darigold Farms. The team names may sound funny, but the league plays serious, hard-nosed basketball. I go up against current college players, former and future All-Americans, and several future NBA players. My former College of Idaho teammate R.C. Owens settles in Seattle briefly and joins the Buchan Bakers, who beat us by a couple of points for the league championship, in a brawl disguised as a basketball game. Throughout the season, I constantly have to fight and claw for position, and I'm often double- and triple-teamed. I still average 33 points and 19 rebounds.

After one game in which I score 40, one of my teammates says, "Someone let the Rabbit *loose*."

I smile, too shy and too embarrassed to respond. But I am happy that my nickname has traveled all the way across the country.

Loading up the car with Johnny after that game, I say, "You're a good coach, Johnny, seriously."

"Well, I have a strategy."

"What is it?"

"Two things: Make sure I put air in the ball. And make sure I get you to the game."

•   •   •

It takes me only a short time to settle into life in Seattle. I love the city. It rains often, but I don't mind, because the rain washes the streets, keeping them clean. I like things clean, even if we're talking about city streets.

I familiarize myself with the city by walking everywhere. I quickly find my "places" — a steakhouse that serves inexpensive and delicious steaks, a couple of nightclubs where top black musicians and singers like Ray Charles, Quincy Jones, and Billy Eckstine play, and a record store that sells my kind of music. I practically live at that record store. As soon

as I hear a song I like on the radio, I buy the single, and when I can afford
it I purchase the album. I accumulate a massive collection of music.

I know all this may sound strange. I'm sitting out a year, biding my
time until I start my sophomore year at Seattle University, and my major activity is playing basketball for Westside Ford. How can I afford to
eat at steakhouses, go to clubs, and buy tons of records?

Well, I have a job.

Ralph Malone put in a good word and I got hired to work in the
men's department at Nordstrom.

It's not a difficult job. I fold shirts and sweaters, stack them on ta-
bles, organize ties and belts, and keep all the clothes neat and organized.
Mainly, though, I stand around and talk to customers. I never have
to hard-sell, because the clothes sell themselves. All I have to do is say
something encouraging, like "Oh, that suit looks great on you," and the
next thing I know I'm ringing up a sale.

One time, a well-dressed man rushes into the men's department and
nearly slams into me. He's breathing hard and looks like he's in a real
hurry. "I'm sorry," he says. "I need a sweater — fast. I'm meeting my
brother in ten minutes and it's his birthday. Only restriction: he hates
red and green."

"Got it," I say. "Right this way."

I lead him to a pile of sweaters on a table. I gesture toward them and
smile.

The guy looks confused. "These are green."

"Oh, right. Sorry. You said no green. My mistake. Right over here."

I move quickly to another pile of sweaters. The guy stares and then
laughs. "Okay. What's the joke?"

"I'm sorry. I don't know what you mean."

"These are red," he says.

Okay. Confession: I'm color-blind. Red and green look the same to
me.

When I drive, I know the red traffic light is the one on top, and
green is on the bottom. But in the Nordstrom men's department, I can't
tell the red sweaters from the green sweaters. It's easy enough to fake it,

though, and I usually can. In this case, the customer finds a stylish blue crewneck and buys it. I ring it up and wrap it—no harm, no foul.

I honestly don't consider being color-blind that big a deal. I don't feel like I'm missing anything, because I've never known any other way. I just have to be careful when I get dressed. I want to be sure that what I wear matches, or at least doesn't clash. So when I buy clothes, I stick to my favorite color, blue, and avoid fancy patterns. Don't want to look like a clown.

· · ·

One Saturday night, I get arrested.

I'm hanging out with Bill Wright, a Westside Ford teammate and a tremendous golfer. Bill will eventually abandon basketball and become the first black golfer to win a United States Golf Association title, on his way to an illustrious career.

But this night Bill and I, acting far less than illustrious, get into a passionate conversation about cars, driving, and drag racing.

Leading to a discussion of who is the better driver.

Meaning the faster driver.

We decide to settle it right then, on the streets.

Bill has a car—nothing special, a family hand-me-down his father gave him. I have a car I borrowed from a friend for the night—a rattling, loud four-door clunker.

Bill laughs when he sees his competition. "You can't beat me in *that*."

"I was going to say, 'You want to bet?' but I don't want to take your money. It would be like stealing."

Bill's laugh dissolves. "It's not about money," he says.

"I know," I say.

We both know it's about winning.

We climb into our respective cars. We establish a finish line a few blocks down Aurora Avenue. We check the area for traffic: all clear. It's after midnight and downtown is deserted.

Somebody counts down and waves a towel, and we peel out. I floor the clunker right away. I honestly don't expect to beat Bill, because my car has zero pickup and an iffy engine, but somehow I stay right with him. I keep my foot pressed to the floor. The engine groans, smokes. I can feel that my clunker is about to conk out. Bill pulls away.

That's when I hear the siren and see the red and blue lights flashing in my rearview mirror. The police car passes me and pulls Bill over. I could easily keep going and the cop would nab only Bill, but I won't do that. We're in this together. I ease off the gas and stop the clunker down the road from Bill and the police officer. I get out of the car and tell the police officer that he needs to take us both in. Bill looks sick. I've heard that his dad can be strict. I can only imagine how he will react to his son's getting arrested. (He won't react well, and will take Bill's car away for months.)

The officer brings Bill and me down to the police station, writes us up for street racing and speeding, and says we can either plead guilty and pay a steep fine or appear in court. I convince Bill to fight the charges. A few weeks later, we get called into court to face the judge. He refers to his notes and says the police officer clocked us going eighty-five miles an hour.

"Your Honor, that can't be right," I say, looking at him incredulously. The judge peers at me over the tip of his glasses.

"There is no way my clunker of a car could go eighty-five miles an hour," I say in the hushed courtroom. "If it could, I'll pay the fine right now."

The judge cracks up.

Maybe he knows Bill or maybe he's heard about me, but he lets us go with a warning. "From this moment on, you two better stick to the speed limit," he says. We do; we never race again. But we don't stop arguing. I know I would've beaten Wright if I had a better car. "It's not the car, it's the driver," he says. "Same thing in horse racing: it's not the horse, it's the jockey."

"It's *all* about the horse," I argue. "Put Eddie Arcaro on Seabiscuit and he'll win every race. But put Eddie on a mule and see where he comes in."

Point: me.

• • •

March 1956.

Our team, the Seattle University Chieftains, makes the NCAA tournament, despite a so-so 18–11 record. Yes, I call it *our* team. Even though I won't enroll in school and join the basketball team until September, the players and Coach Brightman treat me as if I'm already on the team. I play pickup games with the guys, and the players and members of the athletic department challenge me to games of H-O-R-S-E. I don't want to brag, but I never lose. To add a little drama, I once play a game where I take all my shots from the stands. I still win. This day in March, the players, the coach, and I ride the team bus from Seattle to Corvallis, Oregon, a four-and-a-half-hour trek, where we've drawn Idaho State University in the quarterfinals. I'm excited about this trip for two reasons:

I really want to support the guys I consider my teammates.

And I'm going to meet Bill Russell.

Russell, the best player in the country, will be leading his undefeated San Francisco University Dons against UCLA in the semifinals, having earned a bye in the quarterfinals because of their record.

I'm going to meet Bill Russell, bond with him, and pick his brain, then I'll watch him play and study him.

One small problem: I haven't figured out *how* I'm going to meet him. Working on that.

I also need a ticket to watch San Francisco play UCLA the next night. Don't have a ticket.

I plan on asking Bill Russell if he can get me a ticket. As soon as I figure out how I'm going to meet Bill Russell.

We check into our hotel in Corvallis. As we stand in the lobby to get our room assignments—I'm staying with Lloyd Murphy and Francis

Saunders—I see a couple of guys walk through wearing University of San Francisco warm-up jackets.

Bingo.

Bill Russell must be staying here, in our hotel.

The next day, we play our game against Idaho State. Turns out to be a tough matchup: dozens of lead changes, a back-and-forth battle. We pull it out at the end, 68–66. Seems like the whole team exhales at once. We have survived. We stick around and watch a mediocre Utah team win their quarterfinal game. Tomorrow, after San Francisco takes on UCLA, we will play Utah. We should beat them, no problem. And San Francisco will definitely beat UCLA.

We could play San Francisco. We could play Bill Russell.

Now I *have* to meet him.

That night, I pace around in the hotel room. I need a plan. I can't think of anything. I feel nervous. I need backup.

I ask Lloyd and Francis to go with me.

"Go where?" Francis says.

"To meet Bill Russell," I say.

"How?" Francis says.

"I don't know. Let's go to the front desk."

In the lobby, with Lloyd and Francis standing behind me, I sidestep a bellman pushing a luggage cart and stride up to the front desk. I flash a confident smile.

"May I help you?" the front desk clerk says, matching my smile.

"Yes, sir. I'm supposed to meet Bill Russell in his room. Unfortunately, I forgot his room number."

The desk clerk's smile disappears. "I can't give out that information."

Just then, the bellman approaches. "Wait, are you . . . *Rabbit?*"

I turn to him. "Yeah."

"I knew it," the bellman says.

"Rabbit?" the desk clerk says.

"This is *Elgin Baylor.* He and Bill Russell are the two best players in the *country.*" The bellman fumbles in his shirt pocket and produces a crumpled luggage tag. "Would you sign this?"

"Sure," I say, grinning at Lloyd and Francis, who look like they're in shock. The front desk clerk offers me a pen. I scribble my name on the bellman's luggage tag. "There you go."

"Thank you," the bellman says. "I read all the magazines. You scored 53 points in your last game at College of Idaho."

"That's right," I say.

"Seven eighteen," the front desk clerk says.

I look at him. "Was that what I scored for the season?"

"No," he says, his smile returning. "That's Bill Russell's room number."

Standing in the hallway outside Russell's door, flanked by Lloyd and Francis, I clear my throat, hesitate, and then knock twice.

"Who is it?" comes the voice of Bill Russell.

"It's . . . Elgin Baylor."

"Rabbit?"

The door swings open and there's Bill Russell. About four inches taller than me, wiry, his eyes dancing, his face lit up in an electric smile, he waves us into his room. "Come in, come in," he says. Once inside his room, his hands resting on his hips, he considers me, sizes me up, and then laughs, an easy, loud, high-pitched, contagious cackle that I will experience a thousand times in my lifetime.

"Rabbit," he repeats. "Pleasure to meet you."

We shake hands.

"And these gentlemen?"

I introduce Lloyd and Francis. He tells us to take a seat. I hesitate, tell him that we don't want to impose, but that we want to see his semifinals game against UCLA and, unfortunately, we don't have tickets.

"I have plenty of tickets," he says. "How many do you need?"

"Well," I say, looking at Lloyd and Francis and then picturing the faces of my teammates, "the three of us and—"

"Here." He picks up a stack from his dresser and presses a dozen tickets into my hand. "This enough?"

"Oh, yeah, thank you, but are you sure you don't need these?"

"Don't worry. I have plenty more."

He offers us drinks, and we each take a soda. We make small talk, and after a while Lloyd and Francis thank him again for the tickets, stand, and excuse themselves, saying they need to get some sleep for tomorrow's game. I stand to go, too, assuming Bill needs to get to bed as well.

"You don't have to go on my account," he says. "Stay for a while."

"What about your game tomorrow?"

"No competition," he says. "I don't need that much rest."

I must look stunned, because he says, "I'm not putting down the other team. I just don't worry about them. They have the problem — they're playing against me."

He pauses, then explodes into that cackle.

I tell Bill I know what he means. I never worry about who I'm playing, either. I just never think about it. I've also never told anybody that this is how I feel.

"Not too many people would understand," he says.

Bill and I talk for hours. We talk about everything.

We talk about basketball, of course.

We talk about college, how we both ended up at small Jesuit schools, the larger, more established universities ignoring us.

We talk about our upbringing.

We talk about racism.

He describes growing up in a segregated town in Louisiana and the bigotry his family experienced. He tells me about the time his father stopped to get gas at a rural service station, and the white attendant refused to give him gas until he took care of all the white customers, even though his father got there first. Furious, his father started to leave. The gas station attendant shoved a shotgun in his father's face and told him to wait until he finished serving everyone else. I tell him about the racism I lived with every day in D.C. I tell him about the time white police officers called my father "boy." I don't tell how they forced him to beat Columbia.

We talk, but we do more than talk. We *connect* — as basketball players, as college students, as young men, and as young black men. I find Bill funny, quirky, intelligent, and sensitive. At one point, I glance at the alarm clock across the room on his nightstand. I can't believe the time: it's past three o'clock in the morning. I get to my feet, feeling light-headed and energized at the same time. I've never spoken to anyone the way I have to Bill, and I tell him so.

"I have a feeling we'll be having a lot more of these conversations," Bill says, grasping my hand.

He's right. We will have many more conversations.

"You got a game in a few hours," I say. "You'd better get some sleep."

"Told you: don't need much against this team."

His eyes glisten, he waits, and then he bursts into his laugh.

The next day, sitting at midcourt, I watch Bill Russell and his San Francisco Dons dismantle a decent UCLA team, 72–61.

• • •

Following the San Francisco–UCLA game, our Seattle University Chieftains take on Utah, a team we should beat. Maybe the guys are looking ahead at the prospect of a matchup against Bill Russell, or maybe everyone on the team has an off day at the same time, but Utah smacks us, 81–72. In the locker room afterward, the players sit on benches, everyone's head down. The seniors fight back tears. Coach Brightman stands before them, but he seems unable to speak. Except for the consolation game we will play the next day against UCLA, the other losing team in our bracket, the season has ended. We've done well: we won nineteen games and made it to the quarterfinals of the NCAA tournament. The team has nothing to be ashamed of. But the players know they have underachieved. I can tell that some of them feel they have failed. And I can see that Coach Brightman is furious.

I want to say something to the players, but at that moment I realize that even though they have welcomed me and included me, I'm not really one of them. I have never played with them. I don't know what they feel. I can't know. I again look at Coach Brightman, who mumbles a litany of clichés, thanks the team, and storms out of the locker room.

The next day, at the game against UCLA, a game that matters not at all, Al Brightman shows up drunk. He lurches when he walks, alternately screams and sulks, and when he sits on the bench I'm afraid that he'll fall over. Someone will later tell me that when you got close to him, he smelled like a brewery.

The game is a joke. The players seem to be going through the motions. I don't think they want to be here. UCLA blows us out, 94–70. I can't watch. I leave before the game ends. Late in the game, or maybe after it's ended, Coach Brightman jumps off the bench, gets into the UCLA coach's face, screams at him, swears at him, and, according to some people, bumps him, possibly even pushes him. The Seattle University athletic director, sitting close by, rushes over and pulls Brightman away. Later I'll hear that the UCLA coach, John Wooden, has issued a complaint to the NCAA. Shortly after the NCAA tournament, Al Brightman resigns as the Seattle University basketball coach.

I never play for him.

. . .

During a pickup game that spring, I see a young guy drift into the gym, lean against the back wall, and watch us. I assume he's a graduate student on his way to the library or a lab. As it turns out, he's our new coach, John Castellani, a twenty-nine-year-old bachelor and former Notre Dame assistant coach. Later John will tell me that he took the job at Seattle University sight unseen, without visiting the school or the city. A hot coaching prospect, he had come in second for two other head coaching jobs, so when the Seattle offer came, he grabbed it.

"I didn't know about you," he'll tell me years later. "Had no idea who you were. I thought your name was L. G. Bailey. But when I saw you play that day in the gym, I said to myself, *I got my All-American.*"

I go home to D.C. for the summer. I play and coach for the Stonewalls, to keep in basketball shape, and, surprising myself, I spend many Friday and Saturday evenings at home, sitting with my father, watching the Friday-night fights and professional wrestling on his new coin-operated TV. I know we will never have a warm, loving relationship the way I do with my mother and my sisters, but, without ever speaking about it, the strain between us lessens. Grudgingly, I accept who he is and who we are. We share those nights in a kind of resignation and calm — even comfort — as my father cheers on this boxer or that and warns this wrestler or that to "watch out."

That summer, during commercials, my father announces two startling pieces of news. He tells me that this season, he wants to see me play basketball for Seattle University. This is the first time he has ever acknowledged that my basketball playing may be something more than mere foolishness.

And then he tells me that he and my mother have decided to sell the house. They want to scale down, move to a smaller place.

I'm losing my childhood home. I'm twenty-one and, even though I don't live here anymore, I feel displaced. And then quickly, I deal with it, the way I do, allowing my emotions to settle in, simmer, and flit away. I don't want to be touched by loss. I don't want to *see* this as loss. So I detach. I lose myself in the grainy black-and-white boxing match before me, two brawlers battering each other and then desperately clinging to each other so neither will fall. The move seems right for my parents. As for me, I've come to the end of an era. It's time for a new beginning. I'm going back to Seattle to start over.

. . .

I move into a dorm on campus, Xavier Hall, sharing a room with Francis Saunders. We begin basketball practices soon after the school year starts, getting ready for our first game in late November. Coach Castellani is a character — loud, fast-talking, energetic bordering on hyper, and surprisingly strict. Some of the guys who played under Coach Brightman don't love his style. Brightman, they say, was looser, easygoing, treated them almost as equals. Castellani, a product of Catholic schools, believes in discipline and tolerates no nonsense. A hard-ass, someone calls him.

I like him. We're polar opposites, and that may be why. He doesn't appear to have an "off" switch. He goes nonstop at a hundred miles an hour. I'm not sure he sleeps. I respect him as a coach and I do what he says. I also tend to clam up around him. Well, I tend to be quiet around most people. Some guys on the team call me painfully shy. A few others go further and say I'm withdrawn. They make it sound like I have some kind of condition. I don't know. I just don't talk that much. I'd rather *show*, not tell.

Even though John is a head coach for the first time and sometimes I think he's stumbling to find his way, he lets me play my game. He allows me to improvise and doesn't confine me, doesn't hold me back. He sees that I want the other guys to do well. I don't showboat. I see the difference, and so does he.

We open the 1956–57 season against the University of Denver, in a tournament at Idaho State University, in Pocatello. For me, tip-off cannot come soon enough. I have not played in an official collegiate basketball game in more than a year and a half. I'm beyond anxious.

I'm so hyped up, I miss my first three shots. I glance at John, who gestures with his palms, urging me to relax. I exhale slowly, calm myself, and then go to work. I crash the boards, corral a rebound, race upcourt, and bank in a twisting layup.

It begins.

I finish that game—my first as a Seattle University Chieftain—scoring 40 points and grabbing 18 rebounds. We win 67–60.

After that, I go on a tear and so does the team. In December, we play another tournament, in Oklahoma City. I make the all-tournament team and win MVP, scoring 80 points over the three games. The new year arrives and we play a home-and-home series with the University of Portland. The first game, I pop in 33 points and grab 21 rebounds. The second game, I beat that, scoring 41 and pulling down 23 rebounds. We play Gonzaga in January and I score 44 and haul in 17 rebounds.

We rip off 18 wins in a row. We end the season against Portland, playing them for the third time. Poor Portland. Nothing personal, but I just abuse them. This time I score 51 and tear down 20 rebounds.

We end the season at 24–3, ranked number five in the final AP poll, and await our invitation to the NCAA tournament—a lock, it seems to me. But university officials decide to go a different way. They accept a bid to the National Invitation Tournament, which will be held in New York City at Madison Square Garden, to many *the* basketball mecca. Nothing wrong with the NIT—it's the second-most-prestigious post-season tournament—but I wonder why we're going there instead of taking our chances in the big dance, the NCAA. I get no definitive or satisfying answer, but rumors fly.

We think they can actually win the NIT, I hear. *They got no chance in the NCAA.*

Well, I think, but say to no one, *I believe we can win the NCAA.*

. . .

March 1957. Madison Square Garden.

Because of our record and high ranking, we bypass round one and go right into the quarterfinals, drawing St. Bonaventure University, who won their first-round game.

From the moment we arrive in New York, everything seems off. Beginning with my parents.

For the first time, they're coming to one of my games together. They take the train from D.C. and I meet them before the game. My mother wears a snazzy topcoat and looks great. My father, here to see me play for the first time, goes all out. He wears a sporty double-breasted suit and a permanent smile. He doesn't say the word, but he looks *proud*.

Going into the locker room with the team, I feel a mood shift. The whole team goes uncharacteristically quiet. Even John, usually a ball of fire, seems subdued. We dress in silence, and I try to identify what I'm feeling. Nerves, I decide. We're *nervous*. I clap my hands. "It's just a game," I say. "Just another game."

Just another game in front of more than ten thousand people.

We jog onto the Madison Square Garden floor for warm-ups, and I feel the tension. The game starts and everyone is out of sync. After I throw a pass over a teammate's head and out of bounds, St. Bonaventure goes on defense and instantly blankets us with a suffocating, trapping press. We turn the ball over again. After they score, they go back into their press and steal the ball *again*. John calls a timeout. We've never faced a defense like this before, and John, frantically scribbling on his clipboard, tries to show us how to break the press — on paper, on the fly. I'm not sure any of us understands. I take the inbounds pass and start to dribble upcourt. I run right into a defender and knock him over, and the ref calls me for charging.

And so it goes. A living nightmare. We barely score in the first half. We come out for the second half and play only marginally better. I turn the ball over a ton and commit a couple of dumb fouls, eventually fouling out with five minutes left in the game. I score 23 points and take down 25 rebounds, but we lose 85–68, getting bounced out of the tournament our university officials thought we would win.

Back in the locker room, the game and the season over, a grim silence enveloping us like a fog, the team showers, dresses, and files out. I feel numb. I later admit to a magazine reporter that this is the worst game I've ever played.

What's even worse is that I've played my worst game in front of my parents.

In front of my father.

The first game he's ever seen me play.

I find them after the game, and I can't believe their reaction. My mother hugs me for what feels like a full minute, holding me, rocking me, tears of happiness streaming down her cheeks. My father pumps my hand, that smile still carved into his face.

"I'm sorry," I say. "We didn't play well. I didn't play well."

"Yes, you did," my father says. "You were great."

I look into his eyes. He means it.

A parent's pride, I realize then, is unshakable. You don't have to be a superstar for your parents to be proud of you.

All you have to do is play the game.

• • •

Summer 1957.

Showdown.

The Big Dipper wants me.

Wilt Chamberlain, a rangy, seven-foot-two giant, unfolds himself from the driver's seat of his sparkling red-and-white Oldsmobile convertible, the size of a boat, and strides toward me like a gunslinger. Shielding my eyes from the midafternoon sun, I wait near the free throw line at Kelly Miller playground. Wall-to-wall people pack the playground, encircling us, forming a fence of humanity. Someone does an informal count and estimates that two thousand people have come here.

Two thousand people to watch a pickup game?

I doubt that.

No—fact, someone else says. *You got Wilt driving all the way from Kansas to challenge Elgin. That two thousand count may be low.*

Just behind me, Francis Saunders and W.W. stand to my left, and Gary Mays, the Bandit—shirt off, right biceps flexed, left nub dangling—stands to my right. Wilt walks with Dave Harris, a former football star at Cardozo High, in D.C., who went to Kansas University and became Wilt's fraternity brother and close friend. It was Harris who set all this up. He didn't have to plan much. He just said the word and word traveled. Word is Wilt will be staying in town for a week or so with Harris and he wants to play me and my guys. Wilt says he doesn't care who plays on his team. He scoffs when somebody tells him that I know the players around here and he doesn't. He doesn't care who I put on my team, he says; he just wants to play against me. He's curious about the D.C. street game. He wants to see how it compares with the Philly game he knows, which he touts as the best in the nation.

We'll see.

Bottom line, Wilt's landed here for a week and wants to play some ball. He wants a showdown. He's serious. Well, I am, too. We all are down here. I'll be damned if I'm gonna let Wilt Chamberlain or anybody else come to my *home* court and claim some kind of title. That is not gonna happen.

Of course, I know Wilt. I should say I know *of* Wilt. After we got drubbed by St. Bonaventure in the NIT, I turned my attention to the NCAA. I read about every game and watched every game they televised. I expected San Francisco to repeat as NCAA champs, but Kansas and Wilt took them out. Then Kansas faced North Carolina in the final and, shockingly, lost in triple overtime. Racism played an ugly part in that: white fans screamed racist taunts from the stands, and that unraveled Wilt. I sympathize with that. How can you not? But then I hear that Wilt complained about the strategy the North Carolina coach used, especially when he instructed his team to double- and triple-team Wilt. Wilt called that tactic unfair, or words to that effect. I laugh when I hear that. He's seven-two, the strongest man on the planet, and he's crying about being *double-teamed?*

Of course, I also know that for Wilt, playing me is extremely personal. He's won all sorts of awards, including Most Outstanding Player in the NCAA tournament and first-team All-America. I made second-team All-America. I averaged 29.7 points per game and finished third in the country in scoring. He finished fourth—behind me. I led the nation in rebounding. He finished fourth—well behind me.

As I've said, I don't really care about stats. But Wilt does.

And this year, for the first time, I've gotten some mainstream press attention. Never had anything like it before. Magazines and newspapers such as *Sport, Sports Illustrated,* and the *New York Daily News* write articles about me, saying unbelievably flattering things. Before the NIT, a college coach on the West Coast told the *Daily News* writer that I was "absolutely the greatest, the best I've ever seen." He went on to say that he would take me over Wilt. My mother clipped that out and pasted it in her scrapbook. I'm pretty sure Wilt's mom didn't.

As Wilt walks toward me, the group massed near us starts to stir. People shout and whoop, and then a roar rumbles through the whole crowd. I can feel the asphalt vibrate. I have to smile. These are my people. The air crackles with anticipation and excitement, the way you feel at a concert when the lights flicker and dim before your favorite performer struts onstage.

Wilt stops in front of me. He doesn't smile. He sort of sneers and sizes me up. We don't shake hands. "I'm going to light you up," he says.

"You must have a book of matches," I say. "Because that's the only way you're lighting up anything here."

We divide into teams. I've already chosen my guys—Francis, the Bandit, Lloyd, and Willie Jones, a good baller and better trash talker. Dave helps Wilt decide on his team. I don't care who he picks. I don't even pay attention; doesn't matter to me. Bill Russell's words echo in my head: I don't worry about my opponents. They need to worry about me.

We go over the rules. No center jump. As a courtesy, since Wilt's the visitor, I let his team take the ball out. We play full-court. Game's to 30 and you have to win by two baskets. No refs—you call your own fouls. You don't shoot foul shots; you take the ball out. But unless you're bleeding, you best not call anything. Nobody ever calls traveling or three seconds, although we're all over an obvious double dribble. Winning team stays and plays the next game, losing team walks, replaced by another five, the team that has previously called "Next." I don't plan to give up the court. I don't plan to lose. I've never called "Next" in my life. Never had to. Not going to start now.

Wilt's team takes the ball out and we begin. I can feel the electricity from the crowd humming through the playground, through the whole *block*. Wilt establishes himself immediately. He's freakishly athletic, agile and fast. He camps out on the block beneath the hoop and dazzles with dunks, finger rolls, and a short fadeaway jump shot he loves to bank off the backboard. But he seems determined to make the game all about *him*. He doesn't really try to get his teammates involved—he completely ignores some of them—and, as tall as he is, he wows with spectacular plays but doesn't dominate the game. We got guys here who can *play*. I got some rabbit hops myself, and I know I'm more versatile than Wilt. And, I admit it, I put on a little show for the folks who have come out, shaking and baking along the baseline, hitting teammates with an assortment of no-look passes, dribbling behind my back and between my legs like Marques Haynes of the Harlem Globetrotters (ironically, Wilt will play on the Globetrotters for a year; I will turn them down twice), and closing the show with a couple of high-flying dunks and then, the curtain call, drilling in a couple of *long*-range jumpers.

The whole time, Willie Jones does not stop talking. He trash-talks Wilt, dares him to move his big ass away from underneath the basket and come outside and guard him. Willie doesn't trash-talk a lot. He trash-talks *constantly*. When he gets tired of trash-talking Wilt, he trash-talks *me*. And I'm on his team! He lobs a high pass toward me. I jump, grab the ball in midair, and slam it through the hoop in a tomahawk dunk that causes the crowd to howl and scream.

"Hey! Rabbit! That ball was going in!"

"Oh. I thought it was a pass, Willie."

"Pass my ass! That was a *shot*. Leave my shots alone or I'll kick your ass after I kick Wilt's!"

We win that first game easily and hold court. Four more guys come forward, they choose Wilt for their fifth, and we run it back.

Our team wins again.

We play a third time, against Wilt and four different guys.

We win again.

We win the next game, too.

We win every game.

I beat Wilt's team every time.

At one point, Wilt mutters, "I'm going up to Philly and bring back my Philly guys to play you."

"Go ahead," I say. "Bring anybody you want. I'll be right here."

Of course, he never does.

I think he likes it here. Wilt stays in D.C. for three weeks instead of one. Dave Harris is a great host, and apparently Wilt is a great guest. Driving in that snazzy red-and-white Olds convertible, Dave takes Wilt to see the White House, the Lincoln Memorial, the Smithsonian, and the Capitol Building. On Friday and Saturday nights, party nights, Dave brings him around to the best house parties and introduces him to several available and interested young women. I don't know how those dates go, but I tell W.W. and Francis, "I hope he scores better at the parties than he does at the playgrounds."

During those three weeks, we play all over D.C. We take our pickup games to Brown Junior High, which is right near Spingarn, my old high school, to Randall Playground, and to Lincoln Recreation Center. We never advertise, never announce where we will play, but every time we pull into a park, playground, or gym, we find two to three thousand people jammed together, waiting to watch us play. These games become an event, a festival. We play dozens of games, almost every day for three weeks.

Wilt doesn't win one game.

Not one.

•   •   •

I don't know if Wilt gets lucky at one of those house parties, but I do. I meet Ruby Saunders, a couple of years younger, a student at Howard University, who is smart, funny, pretty, and half my size. I wonder why I've never seen her before.

"Where'd you go to high school?" I ask her.

"Dunbar."

"That figures. I should've known."

"Why?" she says, laughing.

"All the good-looking girls go to Dunbar," I say.

I smile. She blushes.

"I went to Spingarn," I say. "I'm Elgin." I wait a beat, expecting her to acknowledge me. Wilt and I have been playing so many games at so many places in front of so many people that I assume she knows who I am.

"Spingarn," she says. "Okay."

"I played basketball," I say, thinking, *Man, I sound dumb.*

"I can see that," she says. "You're tall. Did you make varsity?"

She has no idea who I am.

I'm liking her better and better.

I smile again. She blushes again.

We start dating, which during this time, the summer of 1957, means taking a lot of walks and going to house parties. But while Ruby likes to party, she also likes to read and actually enjoys going to school. She has her mind set on graduating college and becoming a teacher. By midsummer, we are an item. By summer's end, we get serious. By New Year's Eve, we get engaged.

• • •

Fall of 1957, my junior year at Seattle University.

John Castellani sets our starting lineup: sophomore center Don Ogorek, who tore it up on the freshman team, averaging nearly 26 points a game; guards Jim Harney, who can pass and score, and Charlie Brown, who can pass, score, and crash the boards, too; and high-energy Jerry Frizzell, at forward opposite me. John comes into the season full of fire. He wants us to run as much as possible and play a ferocious man-to-man defense. He has his eye on making a deep run into the NCAA tournament. We talk about this sitting in his office, reflecting on last year, both of us believing that we fell short of our potential with our first-round loss in the NIT. We want redemption. To that end, John arranges a tough nonleague schedule. He wants us to face the level of competition that we will play in the tournament. He wants us battle-tested.

In December, we travel to Illinois for a tournament and face last year's NIT champion, Bradley University. I score 28 and we squeak out a hard-fought 82–76 win. Over Christmas, we return to New York City and Madison Square Garden for the annual Holiday Festival. We open the tournament against the University of Connecticut, another hard-nosed East Coast team. We exchange leads all night. The game goes down to the wire, and I make two free throws to seal the 87–83 victory. I score 25 and rip down 22 rebounds.

The next night, we take on Temple University, led by their All-American guard, the slick ball handler and high scorer Guy Rodg-

ers. I know he'll get his points, so I decide to use my advantage on the
boards to gather as many rebounds as I can. *The team that rebounds better
usually wins,* I remind myself. I rebound with abandon.

At one point early in the second half, I jump for a rebound, yank the
ball off the backboard, and get cracked in the head by someone's elbow.
I fall backward and land on the floor. I dab my forehead above my right
eye and stare at my two fingertips, which are doused in blood. I get to
my feet, walk unsteadily off the court, and head to the bench. John directs me toward the locker room, where a doctor employed by the arena
walks with me. I'm not cut badly—I don't need stitches—but I feel
lightheaded and wobbly. By the time the doctor patches me up and I re-
enter the game, I'm wearing a bulging bandage above my eye and we're
down by 20. I play the rest of the game in a daze. I score only 18 points
and we lose to Temple, 91–73.

I'm out of sync the next day, too, for the consolation game against
Dayton. We lose by a bunch and I foul out. As Temple takes the tournament and Guy Rodgers wins the Most Valuable Player trophy, I make
myself a promise. I've had enough of losing at Madison Square Garden.
The next time I play here—if there is a next time—I'm going to show
the hoop-crazed patrons of this basketball mecca the real Elgin Baylor.

• • •

I sit in John's office, my head slumped onto his desk. He rolls his chair
against the wall, cups his head with both hands, leans back, and frowns
at the ceiling. "I forget," he says. "Is it 'Feed a cold and starve a fever' or
'Starve a cold and feed a fever'?"

"I don't know," I mumble.

"I would think you *feed* a fever. That makes more sense. I don't think
you should starve under any circumstance, do you?"

I moan and close my eyes. After our New York trip and the start of
the new year, 1958, I began playing with a vengeance. Poor Portland.
Again. I tore them up for 48 points. Tonight, in a few hours, we play a
very physical Montana State team, and I feel feverish and weak.

John pushes off from the wall and glides his chair back toward his desk. He reaches over and rests his hand on my forehead. "You're warm, Edge."

Edge. That's what John calls me.

I think he's actually trying to say "Elg," but John talks so fast that sometimes he garbles his words. Either way, I think he likes the nickname, and I don't mind it, either.

"I need you for Montana State, Edge," he says. "I need you for every game, of course, but these guys are *rough*."

"Don't worry, I'm going to play . . ."

The words trickle away. Somehow I force myself out of my chair, over to my dorm room, and into my bed.

I take a long nap but still feel feverish when we tip off against Montana State. Every player on the team is a lumbering giant.

It's crazy. When I feel sick, I usually play better. I don't know why that is. Maybe knowing I'm under the weather, I overcompensate. I push myself harder and my adrenaline pumps faster.

We crush Montana State. I score 53 points, breaking Johnny O'Brien's school record of 51 — "Just a matter of time," Johnny says — and pull down 22 rebounds. A week or so later, the fever or flu has passed, and, in front of a huge crowd at the Seattle Civic Auditorium, I glide through layup drills as we prepare to play Portland. Yes, poor Portland. I love playing Portland. I show them my love, scoring 60 points and establishing the all-time Seattle University scoring record, which still stands.

Going into February, I'm on fire and having fun. I score 42 against Gonzaga, and then we play a weak Pacific Lutheran team and I score 51 points and snag 37 rebounds. Many years later, someone will say, "Elgin, these are video-game numbers." I guess they are. But when I'm playing, I'm just *playing*. I'm honestly not conscious of how many points or rebounds I've accumulated until I glance at the stat sheet after the game. Sometimes I can't believe the numbers myself.

I finish the season leading the country in rebounds again. I come in second in scoring, averaging 32.5 points per game, a tick behind Oscar Robertson, in his sophomore year at the University of Cincinnati, and again ahead of Wilt. I bet he's thrilled about that. As we await an invitation to the NCAA tournament, a reporter asks John Castellani about all the points I scored and how I finished second in the nation. He tells the reporter that if I cared about stats, I would've gone for the scoring title and easily beaten out Oscar.

We end the season with a record of 23–6, ranked eighteenth in the final AP poll. We get our invitation to the NCAA tournament, but I feel kind of disrespected. The Associated Press believes there are *seventeen* teams better than us? We will have to see about that.

We head to San Francisco's Cow Palace for the first round of the NCAA West Regional. According to what I read in the sports pages, this region is loaded, the two cofavorites being the University of California, coached by Pete Newell, and the University of San Francisco, the perennial West Coast powerhouse, a team that's reached the NCAA championship game three years in a row behind Bill Russell and Coach Phil Woolpert. I scour the papers the day we arrive and find barely a mention of us. We seem to be an afterthought.

We draw the University of Wyoming for our first game. We beat them by 37. I score 26 points—in a little more than a half. I spend most of the second half on the bench next to John, both of us watching the massacre, both of us wanting more.

In our next round, we face the San Francisco Dons in front of more than sixteen thousand people, at the time the largest crowd ever to see a basketball game on the West Coast. I don't follow the oddsmakers, but I've heard that the Dons are coming in as heavy favorites. Some people predict they will do to us what we did to Wyoming.

They're wrong.

We play San Francisco tight from the opening tip.

Right before the half, with time running out, they go up by four. I take the ball out and see Jerry Frizzell streaking downcourt like a wide receiver. For a moment, I imagine myself playing quarterback in some D.C. sandlot football game, but instead of Jerry, I picture my buddy R.C. Owens flying toward the end zone. I rear back and throw a bomb. The ball arcs the length of the court like a missile and lands in Jerry's outstretched hands, hitting him perfectly in stride as he lays the ball into the basket a second before the buzzer sounds.

"You threw that ball ninety feet on the fly," a teammate says as we jog into the locker room, the crowd noise so loud it drowns out my thoughts.

"I just know we cut the lead to two," I say, patting Jerry on the back.

I go into the locker room so jacked up I have to take a walk. I leave for a few minutes, pace in the hallway, and come back in. I take my seat on the bench, my legs pumping, my whole body pulsating, and lean in to hear John's halftime speech. I stand immediately. I want to get back out there and play. I want to take down these heavily favored Dons.

The second half turns into a war. We trade baskets, squander a lead, fight back, and then, with a little over a minute left, I step to the foul line and sink two free throws to give us a one-point lead. We go into our man-to-man defense, and with ten seconds left, the ref blows his whistle, calling a foul on one of our guards. The San Francisco player steps to the foul line for two shots. He misses the first but makes the second, to tie the game at 67. John frantically calls a timeout.

We huddle around him. He squints at the clock, registers that we have ten seconds left in the game. He scrunches his forehead, slaps the clipboard he holds against his thigh, and rakes his hand through his hair. He kneels before us. "Here's the play," he says. "Give it to Elgin."

Jim Harney takes the ball out under the Dons' goal and, being pressured, finally passes it to Charlie Brown, who's being dogged by the Dons' point guard. As Charlie dribbles, I streak for the hoop, brake suddenly, and loop back toward him. Charlie tosses me the ball. I start dribbling and check the clock.

Seven seconds.

I dribble to half-court.

Six seconds.

The entire San Francisco team masses above the key, sealing me off. In my mind I hear Coach Woolpert's hoarse voice screaming at his team above the din of the crowd, *"Do not let him drive!"*

Five seconds.

I fake one way, go the other, counting the seconds down in my head.

Three seconds.

I'm forty feet from the hoop.

I start to drive.

I stop.

Two seconds.

I step back, jump, and shoot.

Swish.

The crowd explodes.

Fans pour onto the court. A mass of people covers every inch of hardwood, surrounding me. A jangle of arms links through my legs and hoists me into the air.

I sway and teeter as a phalanx of fans carries me down the sideline. I wave, I weave, I blink at the scoreboard to make sure:

Seattle 69, USF 67.

We've won.

We survive. We advance.

I ride the wave of wild, bobbing shoulders as I'm held aloft by this army of crazed Seattle University fans for what feels like ten minutes.

And then everything slows. The adrenaline rush wears off and my body slumps in exhaustion. I feel myself being lowered, and my feet land gingerly on the floor. I turn to find my teammates and suddenly I feel lightheaded. The bleachers, the baskets, the faces circle and swim around me. And then something strange happens. Something *pops*. As if my mind is a slide show and we've gone to the next slide. *Wham.* Reality hits. I break into a run, sprint into the tunnel, and rush into the locker room. I join the screamfest in progress. John sees me, grabs me, hugs me, and shouts, "Just like I drew it up, Edge!" Later, dressed, calm, I check the stat sheet and see that I scored 35 points and pulled down 20 rebounds. Pete Newell, the University of California coach, who watched the game a few rows behind our bench, tells reporters that he's just seen the greatest performance of all time.

Maybe he's being kind and complimentary.

Or maybe he's trying to psych me out.

We play them next.

· · ·

Pete Newell, the Cal Bears' coach, has my number.

Three.

As in: he puts three guys on me from the tip-off.

He has delivered a clear message:

I'm not letting Elgin Baylor beat us.

In addition, Newell directs his team to play at a painfully slow pace. Fighting through the triple teams, I manage only a handful of baskets in the first half. We go into the locker room trailing 37–29.

The second half starts out the same. Halfway through the third quarter, we trail 48–41, and John calls a timeout. I'm frustrated, the team's frustrated, but John seems eerily calm. He tells us to slow our pace down — to play as deliberately as Cal — and to pressure them on defense, get up in their faces, and stay there. He wants us to wear these guys like a suit of clothes.

The timeout ends, and the team heads back onto the court. John holds me back. "Set the other guys up, Edge," he says. "Let's beat them at their own game."

We start forcing turnovers. The Cal players, not used to being guarded so closely on defense, start committing silly fouls. We go to the line time after time, and as a team we make something like fourteen free throws in a row. I look for my teammates and find them, especially my buddy and roommate Francis Saunders. He has the game of his life, knocking in 10 points to spark our comeback. With only a few seconds left, down 60–58 and with every Cal player, it seems, guarding me, Charlie Brown maneuvers into an open spot and bangs in a jump shot to tie the game and force overtime.

In our huddle before the five-minute extra period, John surprises us again. "Stall," he says.

We do. We pass the ball back and forth, playing keep-away for something like two minutes. I don't love this tactic—never have—but it works. Finally, on John's nod, I break for the hoop and swoop in for a layup to give us our first lead. We never trail again. With guys all over me, Charlie Brown hits a long jumper to seal the game, 66–62. We beat Coach Newell, a master tactician, at his own game. A week later, we board a prop plane in Chicago and take a terrifying, stomach-churning flight through tornado-racked skies, landing in Louisville, Kentucky, for the Final Four. On the ground, jittery but alive, our eighteenth-ranked Seattle University Chieftains, the huge underdog, prepares to face Kansas State University, the number-one team in the country.

David versus Goliath.

• • •

On their road to the Final Four, Kansas State, featuring a frontcourt wall of giants, all at least six foot eight, led by All-American Bob Boozer, had beaten both Kansas University, with Wilt Chamberlain, during the regular season, and the University of Cincinnati, with Oscar Robertson. To them, we must seem like a breather.

We again slow the pace, dribbling and passing to one another outside, staying away from their forwards. I soon discover that, while each member of the Kansas State front line towers over us, we're much quicker. I control the boards early, and it seems as though every time one of their big guys goes up for a shot, I go up higher and swat it away. We take a big lead at halftime. When our six-five center, Don Ogorek, picks up his fourth foul guarding Boozer, John moves me over to check him.

I shut him down. Boozer has scored 14 easy points in the first half, but I hold him to one basket in the second half. At the start of the fourth quarter, we pull away. I can't even hear the crowd. I know they're out there, but it feels as if all 18,800 of them have gone mute. Nobody predicted that we would even have a chance against Kansas State, much less take them apart. We go up by 20. The game turns into a rout. And then—

I blow by Boozer, who just can't stay with me. Looking for help, he hollers, "Switch!" I go up for a shot, and Boozer, completely turned around, flails his arms in frustration and accidentally rams me in the side with his elbow.

Pain shoots down my left side like an electric shock. For several seconds, I can't breathe. I gasp and bend over, trying to catch my breath. The first person who comes over to me is Bob Boozer. He rests his hand on my shoulder. "I'm sorry, man," he says. "It was an accident. I didn't mean it."

I try to nod, but I can only grimace. I hobble over to the bench, clutching my side. John points behind me, and I walk directly into the locker room. After the game—we win 73–51—a doctor determines that I have a cracked rib. He swathes me in bandages. Later, for the championship game against the University of Kentucky, he fits a cloth contraption around my torso that looks like a girdle. I try it on, groan, and wriggle out of it.

"I'm not wearing this," I say.

"It'll keep that rib in place," the doctor says. "And relieve some of the discomfort."

"Rather play in pain," I say.

• • •

John asks us to watch the other semifinal game, which follows ours: the University of Kentucky against Temple University. He wants us to scout the team we will play for the championship. I don't pay much attention to the players. Can't seem to focus. My mind drifts, and instead I look at the huge crowd, craning my neck as much as my cracked rib will allow.

"The place is packed," I say to John. "How many people you think are in here?"

"Someone told me nineteen thousand," John says.

I whistle. "*Nineteen* thousand people?" I look left, I look right, I look across the court at the stands on the other side. I nod at Charlie Brown and Francis Saunders and turn back to John. "So that would be me, Charlie, Francis, and eighteen thousand nine hundred and ninety-seven white people."

I smile at John. He smiles back at me uncomfortably, probably wondering if I'm joking or not.

I'm not.

Kentucky and Temple battle down to the wire. Kentucky, the beneficiary of several very questionable calls, wins a 61–60 nail-biter.

"Talk about home cooking," I say to Francis as we slowly file out of our row. "Temple got screwed. I hope we don't get those refs."

We get those refs.

We take the court against the Kentucky Wildcats, an all-white team, coached by the legendary Adolph Rupp. Even though my side feels as if I've been stabbed, I try not to let on that I'm feeling any discomfort at all. I can't, for my own protection. If my opponents know that I have a cracked rib on my left side, they will go right *at* my left side. It's the law of battle. If you discover an opponent's weakness, you attack it. Especially when you're fighting for a national championship.

I shake hands with the Kentucky players, trying to swallow the sudden rush of pain that runs down my side. As I move into position for the jump ball, each breath causes me to pant. I crouch in the center circle. The referee tosses up the ball. I jump, tap the ball to Jim Harney, and shut everything out, including the pain. All I see, hear, and feel is this game.

John asks me to guard John Crigler, an active but low-scoring guard who went scoreless against Temple. We trade baskets with Kentucky early. Then Coach Rupp calls a timeout and instructs his team to go into a pick-and-roll offense with Crigler, who either sets a screen for Vernon Hatton or Johnny Cox, their high scorers, or rolls to the basket to receive a pass from one of them and go in for a layup.

The first time they go into their pick-and-roll, I don't bite. I stick with Crigler as he takes a pass from Hatton. When Crigler goes up for a layup, I block the shot cleanly.

The referee blows his whistle, calling me for a foul.

I quickly walk away.

We play to a 10–10 tie, but then we stop the Wildcats three times in a row and I go to work. I hit a jump shot from the top of the key, drive toward the basket, dish off to Francis for an easy lay-in, and then hit a jumper from the left of the foul line. We go up 21–16. On the next Kentucky possession, one of their players goes up for a tip-in off a missed shot. I block the tip-in attempt and scramble for the loose ball.

The ref blows his whistle and signals me for a foul.

I slap the ball between my hands, pause, and hand the ball to the referee.

I hit another jump shot, put in a short floater across the middle, and hit Charlie with a no-look pass on a fast break.

We go up by 11.

Kentucky tries another pick-and-roll off a Crigler screen. I fly by Hatton and catch up to Crigler as he goes up for a shot. I swat the ball away from behind.

The referee whistles me for my third foul.

I race over to him. "Sir, I got all ball. I didn't touch him."

The referee whips around and glares at me. "Shut up and play, boy."

I freeze.

Boy.

Now I burn.

I start to speak, but I smother the words. I bite my trembling lip and walk away.

Eighteen thousand nine hundred and ninety-seven to three.

That's what I see on an imaginary scoreboard in my mind. That's what the score feels like: eighteen thousand nine hundred and ninety-seven white people to three black people.

Charlie, Francis, and me. Especially me.

I can fight this war, a war bigger than basketball, only one way.

I lift my head, find new resolve, and play on—harder.

Crigler, the recipient of at least two layups off that pick-and-roll play, and Hatton and Cox, popping shots from the outside, keep Kentucky close. I match them basket for basket. On defense, I fasten my arms to my sides, not wanting to risk a fourth foul before halftime. The first half ends. We jog into the locker room, clinging to a 39–36 lead.

John doesn't offer much of a halftime speech or pep talk. We all know we're in the midst of a strange game. We should beat this team; we all know that. We've beaten better teams to get here. I refuse to believe that the referees are targeting me. I can't go there. I would certainly never use that—or anything else—as an excuse. But I do feel that on some plays, we're not only playing against the five Kentucky players—we're also playing against the two officials.

The second half starts and we quickly build our lead to 44–38. Then, four minutes into the half, one of the refs calls another phantom foul on me, number four. One more and I foul out. John, racing down the sideline, screams for a timeout.

We squeeze around him as he frantically draws a cluster of stick figures on his clipboard. "Listen up," he says, his voice rising into an alarming screech. "We're going to play a zone."

In the two years he's coached us, we have never played a zone. I'm sure it's not difficult to figure it out, but we'll be playing it *now,* for the first time, in the NCAA championship game. Learning on the fly. Learning by doing.

John points out where I will play, tapping the clipboard with his marker. I'm to guard a spot, not a player. And if it looks as if the player closest to me is about to score, John wants me to concede the basket.

"Don't challenge any shots, just get the rebound," he says. "You can't foul out, Edge. We got almost a whole half left."

I nod, I grumble. I don't like it.

"One more thing," he says. "On offense, slow it down. Go into a stall."

I like that even less.

We give it our all. On defense, we go into a zone, with the other four guys, on top, chasing Kentucky players through their crisp pick-and-roll offense. On offense, we slow it down, killing our momentum. We turn the ball over repeatedly. Kentucky catches us, and with Hatton and Cox hitting jump shots and Crigler cutting to the basket for uncontested layups, they pull away. We lose 84–72, a game we should have won. A game we *had* won.

In the locker room after the game, my side on fire, I sit slumped on the bench, both fuming and exhausted. I've scored 25 points, pulled down 19 rebounds, dished off two assists, and blocked two shots. I don't care. We didn't win.

John charges into the locker room and slams his fist against a locker, causing the metal to scream. He turns and speaks so low I can hardly hear. "It's my fault. I shouldn't have, I *should* have . . . I don't know what I should have done. I got outcoached. Rupp has six hundred wins; I have sixty. Still. We should've won."

And then I hear a rustling in the doorway and Coach Rupp — tall, dark-haired, natty in a brown suit, looking like a statesman — stands for a moment at the end of the row of lockers and then makes his way over to me. I force myself to stand.

"I just want to say congratulations," he says. "We were lucky to beat you. This is as fine a team as we've played all year. And you're a great player. A *great* player. I wish you nothing but the best."

He offers me his hand.

I hesitate for a half second and then we shake.

He walks over to John and shakes hands with him, and then Coach Rupp strides out of the locker room. A few minutes later, in a ceremony at midcourt, a member of the NCAA tournament committee will hand me the trophy for the Most Outstanding Player of the 1958 tournament. But first he presents Coach Rupp with the championship trophy for his Kentucky Wildcats. And then he turns to hand John Castellani the second-place trophy. Except John is not there. He's remained in the locker room, sitting slumped on the bench, his head in his hands, too upset to move.

"We had it," he moans, his words echoing in the empty room. "We should have won. I let them down."

• • •

The chatter starts.

I hear it from John Castellani, from my teammates, from my parents, from my sisters, from my fiancée, Ruby, and even from TV celebrity Steve Allen when I appear on the March 30 national broadcast of *The Steve Allen Show*.

"Elgin, are you going to turn pro?"

I shift uncomfortably on the stage of the theater, near Times Square in New York City, and squint into the blinding bank of floodlights aimed directly into my face. I stand with the other members of the USBWA/*Look* magazine college basketball All-America First Team — Wilt Chamberlain, Oscar Robertson, Guy Rodgers, and Bob Boozer — the first time ever that *Look* has chosen five black first-team All-Americans.

I look at Steve Allen, think about his question, stammer, and say, "I . . . I . . . No. I'm going to stay in college for my senior year." I'm vaguely aware of a sudden burst of applause booming toward me from the studio audience in the darkness below.

I'm not sure why I tell Steve Allen and a national television audience that I'm not turning pro, because I honestly don't know. The words just tumble out, unchecked. The truth is that I feel torn and confused and pressured. It seems I can't turn around without somebody asking that question. And the question doesn't just come from my family, friends, and fiancée.

The Harlem Globetrotters ask *again*.

Someone from the New York Knicks calls to gauge my interest.

Would you like to play for the Knicks, in Madison Square Garden?

Who wouldn't?

Then I hear a rumor that the Knicks have offered the Minneapolis Lakers $100,000 for their first pick in the upcoming NBA draft, which the Knicks want to use to pick me.

But Bob Short trumps them all.

Mr. Short, owner of the hapless Lakers, a team in trouble due to both a shortage of funds and a lack of talent, literally shows up at my door when I come home for a visit over spring break. Apparently, John

Kundla, the coach of the Lakers, attended the Final Four in Louisville, saw me play, and told Mr. Short that, in his opinion, I could turn the team around. Now Bob Short, barely forty, who only recently acquired the Lakers and is a very hands-on owner, sits on our living room couch and not only guarantees to take me with the number-one pick, but promises to pay me at least as much as the Celtics paid Bill Russell: $22,500.

"I can improve on that," he tells me, my parents, and Ruby.

My mother nods, Ruby squeezes my hand, and Pops fights back a smile. The salary Bob Short mentions exceeds any paid to a rookie in the history of the NBA. I catch my father's eye. He nods almost imperceptibly. We've officially come a long way from the days when he referred to my basketball playing as foolishness.

"I don't know," I tell Mr. Short, because I really don't know. I'm confused enough right now to flip a coin. My future in a coin flip. That, of course, would be insane. I feel so tempted.

"You can always finish college in Minnesota," Mr. Short says. He assures me that with one phone call, he can get me into the College of St. Thomas, where both he and Coach Kundla earned their degrees.

"It sounds . . . I don't know," I say. "I'm not sure. I have to think about it."

"Absolutely. You've got until April 21. Three weeks."

We all look at him. Bob Short smiles. "The draft is April 22. Of course, I wouldn't mind a *little* notice . . ."

I return to Seattle to mull over my options. I try to picture myself in my own future. I can see myself here in Seattle, no problem. Staying here feels so comfortable. I love the city, love the people, and I know we would definitely make a serious run at the NCAA title once again. We have a lot of players coming back, including Charlie Brown, who is poised to become a star, and Coach Castellani has been actively recruiting a couple of strong players from Washington, D.C. I've met one of them, Ben Warley, a six-foot-six forward. I've talked to Ben a couple of times and run with him on the playgrounds. Good player. One day John tells me that Ben has given him a verbal commitment, but he's heard that at least one other school is also going after him hard. John says that he'll recruit Ben even harder to make sure he comes to Seattle. I don't know what that means, but I don't like the sound of it. I tell John to watch his back.

"Why? Is there something I should know about this guy?" John says.

"I don't know anything specific, but you hear things." I pause. "I'd be careful."

I take a couple days for myself, walking through campus, strolling through the city, hanging out in my favorite record store, listening to music, trying to distract myself, hoping the decision whether to stay and play college ball or go pro will reveal itself. At one point, I imagine what my life would be like if I didn't play basketball at a high level. I picture myself graduating college and following my most likely career path: becoming a schoolteacher. I would probably end up in a junior high school somewhere, teaching gym classes and coaching the varsity basketball team. As for salary, I'm told that schoolteachers make around $6,000 a year, plus good benefits and summers off. Not a bad life.

Teaching junior high school or playing in the NBA?

I know: the choice seems obvious. But—no pun intended—it's not a slam dunk.

I think about turning pro early. Even though I'm not a senior, I'm eligible because I sat out the year I transferred. I wonder how I would do in the NBA. Am I ready? I'm not worried about making it. Not too worried, anyway. I think I'll do well, but how well? How will I match up against a whole league of quality players like Bill Russell, Wilt, and Oscar? Will I hold my own? What if I turn pro and I'm not good enough to start? How would I adjust to being a role player? I *know* I'll do well if I stay in college. I've already seen that.

Then I think about the practical side. What if I stay in college and suffer a serious injury? That thought always crosses my mind. You can't prevent some freak accident from happening, like taking an elbow in the ribs the way I did in the tournament against Kansas State—or worse. I know guys who've gotten seriously hurt and couldn't play again. I believe my value as a player has reached its peak *now*. I should take advantage of that. Right? Shouldn't I?

I decide to hash this out with John, especially since I'm running out of time. I need to come to a decision and stick with it. I drop in to see him at his office one afternoon, only to be told he's in a meeting behind closed doors. His secretary tells me that he won't be able to see me at all that day. I thank her and leave, feeling something's not right.

When I return to see him a day or so later, he's not there, and he's locked up his office. He's taking a personal day, his secretary tells me. When I go back a third time, I find his secretary gone, her desk cleaned out, and John's office vacant. Panicked, I ask everyone I can find what happened. Eventually I get an answer: the university has terminated John. He has moved back to his hometown in Connecticut.

That night, he calls me. He starts to tell me what happened, then his voice cracks and he stops talking. He tells me he's exhausted and promises to explain everything the following day. After he hangs up, I realize that he got choked up on the phone and couldn't talk. He didn't want me to hear him that upset.

Finally, over the course of a few days — and then over a much longer period of time, years actually — I piece together the details, or at least as many details as I want to know.

John has broken the NCAA rules on recruiting in a big way. He shattered them. He got caught offering money and plane tickets to Ben Warley and another player he recruited, George Finley, a seven-foot center. Warley and Finley had agreed to enroll at Seattle but backed out at the last minute, saying they'd changed their minds and decided on another university. John begged to meet them one more time to plead his case. They reluctantly agreed.

Turned out to be a setup.

John met them at a secret location and offered them the best deal he could — the plane tickets, $200 a month as a "stipend" to Warley, and $90 a month for Finley.

What John didn't know was that one of them was wearing a wire. After the meeting, they turned John in. Not only did the administration

fire him, but the NCAA suspended Seattle University from any postseason NCAA tournament play for two years.

Hearing this—digesting all of it—I feel profoundly sad.

I like John. I love his spirit, his competitive drive, his fiery personality. I know he has only one intention: to win. I don't know this for a
fact—I've never asked John about it; I've chosen not to—but I imagine
that he saw what he did as no worse than what he saw occurring at other
programs. He just wanted to be competitive. I remember him telling
me, "If I get Warley, we win the national championship next year and
maybe for years after."

I hated losing to Kentucky. That loss stung. But I didn't realize how
much it tore John up.

• • •

I decide to turn pro.

I officially enter the 1958 NBA draft.

I don't make my decision because of John, but his leaving makes my
decision a lot easier.

At Pops's suggestion, family friend Curtis Jackson offers to represent
me, or at least advise me. I'm not sure of Curtis's credentials or negotiat-
ing skills—he works for the post office—but I know him to be tough,
savvy, and levelheaded. Most of all, I trust him.

An hour before the draft, I sit in my dorm room with a few friends
and Curtis, who has flown to Seattle from D.C. I face the phone, which
sits a few feet away on my desk. At this time, none of the three television networks televises the draft, ESPN has not yet been invented, and
we can find no radio coverage. Waiting for that phone to ring, I feel
nervous, isolated, and unbelievably vulnerable. Yesterday Curtis spoke
to Bob Short to tell him I was entering the draft and to get confirmation
that the Lakers would draft me with the number-one pick.

"He's made his decision," Curtis told Mr. Short. "Go ahead and draft
him."

"We probably will," Short said.

Wait.

"Probably"?

I panic.

"He's just playing poker," Curtis assures me.

And so I wait, a wreck in my dorm room, my friends trying to break the tension by joking while we play poker ourselves. But I can't concentrate on my cards. I keep glancing over at the phone, trying to will it to ring. I can't help thinking that my future — my entire life — depends on how quickly that phone rings and who will be on the line.

I do expect the Lakers to draft me. I also know that teams trade up at the last second and strange things can happen.

Someone starts dealing the cards. I pick up my hand, fan out my cards, and —

The phone rings.

I dive for the receiver, lift it to my ear, fumble with it, drop it, pick it back up, and practically shout "Hello?"

"Buy a heavy coat," Bob Short says. "You're moving to Minneapolis."

Seeing my expression, my friends shout. I hand the phone and the negotiations to Curtis.

He doesn't have much to negotiate.

Bob Short makes good on his promise. He offers me $25,000, the highest salary ever paid to a rookie in the NBA.

I finish my junior year at Seattle University, pack up my dorm room, and return to D.C. for the summer. I play pickup games every chance I get, determined to stay sharp for the Lakers' training camp, which will begin at summer's end. I never miss a day, playing mornings, afternoons, and evenings. Except on July 10, when I play only in the morning.

In the afternoon, I get married.

My wedding to Ruby takes place at Mount Olive Baptist Church, in front of close to a hundred family and friends. That day — the anticipation, the excitement, the ceremony, the solemnity, the joy, the music, the dancing, the celebration, all the details — will eventually blur in my memory and go out of focus. I'll no longer be able to picture the faces of those who shared that day with us. I'll even have trouble picturing Ruby. The truth is, I tend to vanish emotionally, especially after time passes and when memories sting. I want to keep them at bay, and so I bury them. Some say I repress the memories that hurt. I say I choose to forget. I block things out. It's how I cope.

We'll try to make it work, Ruby and I, but we're also probably getting married for reasons that destine our marriage to fail. My ears have scorched from stories I've heard about the lifestyle of NBA players. Someone told me that the NBA consists of 50 percent money and 50 percent sex. *I don't hear "basketball" in that equation,* I thought. Maybe I've been sheltered, or maybe I'm naive, but all I care about and all I think about is basketball. And so, to protect myself from any sort of distraction, especially on the road, I get married.

Years from now, I'll also realize that I wanted to escape from the bitterness and pain I saw everywhere in D.C. But I act by instinct. I don't consciously know how much of the brutal residual racism I carry or how urgently I want to leave that city behind. Once I leave, though, I'll be gone. Sportswriters will note that fact repeatedly. I almost never come back.

But I won't suppress every moment of my wedding day. I'll always remember my family, my sisters especially, dancing and beaming, and I'll remember the music and the joy that surrounds us. I'll recall Francis Saunders, my best man, perhaps a little tipsy, making a toast to the two of us, hugging us both, trying to hold it together, maybe realizing more clearly than I that I will soon be leaving for good. And later, on the dance floor, with flashbulbs popping, will come that moment when I lift Ruby high in the air so the newest Harlem Globetrotter, Wilt Chamberlain, can kiss the bride.

Instead of taking a honeymoon relaxing on a beach in some exotic locale, we drive from Washington, D.C., to St. Paul, Minnesota, and settle into our apartment in the Twin Cities. Ruby prepares to enter her junior year of college at the University of Minnesota, continuing her dream of becoming a teacher, while I begin what I hope will be a decent career with the Lakers.

6

LAKER

THE LEGENDARY GEORGE MIKAN. ALL-STAR JIM POLLARD. The two architects of five championships in six years for the Minneapolis Lakers, *the* dominant team in professional basketball.

No more.

Those days have gone. Pollard has retired, and Mikan, hired as coach, left the team in the middle of last season. The Lakers finished the 1957–58 season with a record of 19–53, last in their four-team division, twenty-two games behind the league champion, the St. Louis Hawks.

I know what I'm getting into. The Lakers need to rebuild, and Bob Short says he's just paid $25,000 for a new foundation: me.

What I don't know is that for years, the Lakers have been hemorrhaging money. Bob Short, considered by some to be a bold investor and by others to be reckless, has come to the end of his patience. I hear that if the team doesn't turn around this year, Mr. Short will oversee a fire sale. He knows when he has a losing hand, and he will walk away. He won't get his money out, but he will stop the bleeding.

Simply put, the Lakers franchise, Bob Short himself, and both of the Twin Cities are betting on me.

No pressure.

Actually, I don't feel the pressure. I really don't. I can't think about anything as lofty as saving an NBA franchise. I'm thinking about something much more basic: I will walk into the gym for our first practice just hoping I can play with these guys. I sincerely hope I can make it in this league.

I've done my homework. I've memorized the names of all my new teammates and their stats from last year. I've read all about them. Veterans Vern Mikkelsen, at forward, and Larry Foust, at center, lead the team in scoring and rebounding, averaging around 17 points a game and 12 rebounds each. Dick Garmaker, a sweet shooting guard, contributes another 16 points a game. Bob "Slick" Leonard, his running mate, gives us 11 more, and Hot Rod Hundley, an NBA legend primarily *off* the court, known for his wisecracks, fast lifestyle, and behind-the-back passes, adds another eight or so.

Overall, as we warm up for our first practice, I like what I see. These guys can play. I'd be happy to contribute 15 points or so my first year and maybe eight or nine rebounds. That would make for a solid rookie year. I know what Bob Short and John Kundla expect, but I make no promises, especially to myself. I just want to contribute.

By the second week of practice, and especially after a few scrimmages, I discover — and admit only to myself — that I'm actually better than these guys. Playing pickup ball in D.C. with the Stonewalls and AAU ball for Westside Ford, in Seattle, prepared me for the NBA more than I knew. It hardened me, allowed me to develop my skills. I've come in a little bit ahead of the class, and I'm itching for the season to start. I can't wait to match up with high scorers Bob Pettit and Cliff Hagan on the Hawks, Jack Twyman on the Royals, and George Yardley on the Pistons, and to play against Bill Russell, Bob Cousy, and the other Boston Celtics. I'm dying to take them on.

For now, though, we practice every day and get to know one another, both on the court and off. From day one, Hot Rod establishes himself as our leader. He lives life with a swagger, as if he is hosting both the game and the after-party. The very first practice, Hot Rod walks over to me, his grin stretching across his face, and hugs me. "How you doing, baby," he says. "Glad you're on *our* team. You're gonna make me look good, baby."

When we play together, Hot Rod makes it his sole mission to get the ball into my hands so I can score. He has taken a page from Cousy, a magician with his passing and dribbling, and will pass the ball behind his back, over his head, between his legs — any way he can — to feed me the rock. Kundla prefers to start Slick Leonard and Dick Garmaker and have Hot Rod come off the bench, bringing a burst of energy the moment he steps onto the court.

The pro game suits my freelance style of play. Recently, two rules changes have been incorporated that seem geared expressly toward helping me. First, to counteract former Lakers center George Mikan's dominance in the low post, the league increased the width of the free throw lane from six feet to twelve. The bigger area allows me more room to operate near the basket. Second, the league instituted a twenty-four-second shot clock. I love that rule. Now teams can't go into that momentum-killing stall. Nothing more boring than watching — or playing on or against — a team that passes the ball back and forth for ten minutes without taking a shot. I like a fast pace. I thrive in an offense that runs.

Finally, after what seems like endless months of practice and scrimmages, in late October, 1958, I play my first NBA game, against the Cincinnati Royals, at home in front of a large, enthusiastic crowd, among them owner Bob Short, who sits behind our bench within whispering distance of Coach Kundla.

Jim Krebs jumps center and taps the ball to me. I sprint toward the basket, shimmy my shoulders, then blow by my defender and drive in for a layup. The crowd roars. I make three more quick baskets, on my way to scoring 25. We blow out Cincinnati by 20 and I think, *Yeah, I can play this game, no problem.*

Well, I can, but I don't.

I scuffle for the next three games and don't reach 20 points in any of them as our team struggles to a 2–2 record. For some reason, I play very tentatively. I just feel reluctant to take charge. After a lackluster 112–100 loss to the Hawks at home, in which I score 13 points and watch Bob Pettit pour in 37, Hot Rod corners me in the locker room.

"You're playing tight, baby," he says. "It's like you want to make sure we're all involved, we're all happy. You're so nice. Screw that. We want *you* to score. Let it loose. Be yourself, baby. Be *yourself*."

He's right. I have been concerned about my teammates thinking I'm some rookie hotshot and a ball hog. I've got to let that go. The next day, we go into St. Louis to play the tough Hawks. I decide to take Hot Rod's advice. I let myself loose. I score 29 and we beat them 108–101.

I elevate my game—score 26 in one game, 30 the next—and still the team struggles. We win a couple in a row and then drop two straight, including a loss to Syracuse on their home court, even though I pump in 34.

We continue on the road to Boston, to play the mighty Celtics in front of a crazed, foot-stomping crowd in Boston Garden. It'll be my first time going head to head against my friend Bill Russell.

• • •

The day of the game with the Celtics, several hours before tip-off, Bill and I meet in my hotel room. We catch each other up and Bill asks me how I'm enjoying life in the pros.

"It's an adjustment," I tell him. "A lot to learn. Mostly off the court, honestly."

"Oh, *yeah*."

"The team's scuffling a little bit right now. We'll come around. The guys are great, though."

"Hardest part's the travel."

"Yeah, I'm finding that out."

"And . . ."

He stops.

"And?" I say.

"There are . . . certain cities. St. Louis—that's probably the worst. I've had trouble in Boston, too, though not with fans at the Garden, not when I'm playing."

I know exactly what he's talking about. "We're making progress, though," I say. "We got three black players: me, Boo Ellis, and Ed Fleming. Better than two years ago. Better than a year ago."

"Slow progress," Bill says. "Got a long way to go."

We talk some more and then make a plan to meet after the game and go out for dinner. Bill leaves and I take a pregame nap.

Game time arrives. In the dressing room, we change into our blue away uniforms. I peel off my shirt and goose bumps form over my entire body.

"It's freezing in here," I say. "It feels like thirty degrees."

"That's because it is," someone says, and everybody laughs except Boo Ellis and Steve Hamilton, the other two rookies.

I'll later hear that the Celtics turn off the heat in the visitors' locker room in the winter. I believe it. It's so cold in here, my teeth start chattering.

A few minutes later, we jog onto the court and go through our pregame warm-ups. At one point I glance over at the Celtics and catch Bob Cousy making shots from the outside, Tom Heinsohn playing one-on-one with another player, and Bill practicing free throws, sinking one, watching another clank off the front rim. I laugh. *Bill does need to practice free throws,* I think.

I turn back to our side of the court, take a couple of dribbles, and drill in a shot from the top of the key. I check the crowd and watch the people filing in. It's early November, only eight games into the season, and Boston Garden rocks with close to a sold-out house. The atmosphere hums. Feels like a playoff game.

The horn sounds. After player introductions, we gather around Kundla. He shouts out instructions that we try to follow but usually forget, can't hear, tune out, or change according to the way the game plays out. The Celtics wait for us to take our positions for the center jump. Bill comes out last, walking slowly, staring straight ahead, his eyes eerily glassy, his expression dark and focused. The players shake hands. Lowering his head, Bill walks into the center circle. I walk over to him, smile, and extend my hand.

He doesn't take it.

He doesn't look at me.

He swivels away, turning his back to me, and when he turns in my direction, he looks past me, avoiding eye contact.

He refuses to look at me.

You are the enemy, he says without speaking, and then his eyes go almost sinister. *You are nothing. You do not exist.*

I don't want to take Bill's reaction personally, but I do. I want him to acknowledge me. More than that: I want to burn my presence into him —and the rest of the Celtics—like a scar.

I play like I'm possessed.

I score in flurries with an arsenal of shots, many that I make up on the spot. I slip past two defenders who converge on me, dip beneath them, spin toward the hoop, jump, hang in the air, and bank in a layup. The crowd gasps. I brush off every Celtic who guards me. I soar past Heinsohn, bull my way through "Jungle Jim" Loscutoff, shoot over Bill Sharman. I sweep the boards like a broom. Once, dribbling down the lane, I come face to face with Bill, his arms wide, his body a wall of defiance. Although his face wears no expression, I can read his eyes.

Come on, Elgin, bring it. I dare you.

Fair enough.

I take off from the free throw line.

I elevate.

Bill jumps at that same instant. He stretches his hand, his fingers flayed a foot above the rim.

My hand—palming the ball—soars even higher . . . rising . . . rising . . .

And I throw down a dunk—

Over Bill Russell's outstretched hand.

The crowd goes silent. They can't believe it.

Bill can't believe it.

Takes me a second before I believe it.

I turn to see Bill's reaction.

He has wiped any reaction away. He's gone stone-faced. Blank.

I jog nonchalantly back toward our basket, my heart thumping, the crowd now rustling and *ahhing* behind me.

We battle the Celtics to a tie at the end of regulation. With the crowd rocking, the noise in the Garden head-pounding, we go into overtime. Early in the overtime period, the referee blows his whistle and calls a foul on me, my sixth, and I foul out for the first time in an NBA game. As I walk slowly to the bench, my head lowered, my spirit crushed, splinters of crowd noise pelting me, I suddenly become aware of—

Applause.

This full house of Celtics fans, the loudest, harshest, most critical basketball fans in the universe, is cheering me, the enemy. I've scored 36 points and hauled down a passel of rebounds. But more than that, I've left every ounce of myself strewn on that parquet floor. The Celtics fans know it—and now the applause builds and they stand for me.

Okay, I'm convinced. Nine games into my NBA career and I know I can play with anybody.

A few minutes later, the Celtics take the lead and hold it. We lose 116–113. I sit on the bench for a moment, my chest still heaving, and I think about Bill. During the game, I tried to make eye contact with him a dozen different times and he never looked at me. Not once. I toss aside the towel I'm holding and follow my teammates into the locker room, which, miraculously, has reached a normal temperature.

We don't say much to one another. We tend to go quiet after a loss. We've played only nine games, but it seems like more, maybe because we've lost three in a row. I *hate* losing. Losing leaves an actual bitter aftertaste, as if I've swallowed something that doesn't belong in my mouth, like a handful of coins.

Sometimes a loss hangs on me like a weight. I can't shake it. I will click through pictures of the game in my head. I'll think about certain plays while I shower, while I get dressed, while I walk down the street, while I eat. I take really tough losses to bed with me. I stare at the ceiling and rewind certain moments, mistakes I made, bad decisions—a shot I took when I should've passed, a bad pass I made when I should've taken

a shot — a stupid foul, a weak attempt at a steal, lunging for a ball I had
no chance for, swiping at it, missing, coming up with nothing but air
and watching my opponent score. I wonder if I could have done more,
or something different that might have changed the loss to a win. Sometimes a loss keeps me up at night.

But, truthfully, this happens rarely.

Most of the time I let a loss go.

This may sound simplistic, but it's the way I deal. The game has
ended. It's over — nothing I can do about it now. I've given everything
I have. I think the healthiest response to a loss is "We'll get them next
time."

After the game with the Celtics, that loss still smarting, images from
the game still stuck in my head, I wait in the hallway outside the dressing room for Bill Russell. I can't believe that during the entire game,
Bill never once acknowledged me, never smiled, frowned, nodded in
my direction — nothing. He completely ignored me. It was as if I wasn't
there.

I look at my watch. He's late.

Maybe he's not showing. Did I offend him somehow? Did I say
something that pissed him off?

The dressing room door opens and Bill steps out. He sees me and his
face lights up. He approaches and drapes his arm over my shoulder.

"So, where you want to eat?" he says. "I'm starving."

I get it. The battle's over. We're friends again.

• • •

We lose six in a row, finally manage a win, and then lose the next three.
We pull it together and reel off four straight victories. We falter again
and lose four in a row. Then, as Christmas Day approaches, we win two
in a row, including a victory at home against the Philadelphia Warriors
in which I score 33. By now we've played twenty-six games, almost
the equivalent of a full college season, staggering to an 11–15 record. I

think we can do much better, but Bob Short seems ecstatic. Last year the Lakers won nineteen games the entire *season*.

Most nights, I lead our team in scoring and rebounding. Not every night. Some nights, usually on back-to-backs, I'm so tired I'm bleary-eyed and my legs feel as if they're encased in cement. During dinner, after a bad loss to the Celtics in a game we played in Charlotte, North Carolina, Bill Russell and I talk about the game's day-to-day grind. The crushing pace we endure—the travel combined with very few off days—has begun to wear on me. We talk about how to pace ourselves and the need for rest. We talk about how we play a very physical game, especially inside, where fighting for position often resembles hand-to-hand combat. And we talk about the brutal schedule. We don't play back-to-backs; we play back-to-back-to-back-to-back-to-back-to-*backs*. And we do that all season.

Looking over our schedule after our December 20 game against Philadelphia, we play six games in six days, starting with a matchup with the Detroit Pistons in New York, the first half of a doubleheader, with two other teams playing the second game, a common practice in major cities like New York and Boston. We skip one day, December 31, then play New Year's Day and then six of the next seven days—thirteen games in fifteen days. We play a viciously condensed schedule: two divisions of eight teams playing seventy-two regular season games in less than five months, with three of the four teams in each division making the playoffs. It seems as if we're always on the road, traveling not just to the cities that have teams but to cities that might, in the future, have teams. We showcase the NBA in Houston, Dallas, San Francisco, Seattle, and Portland. We also play in cities *near* cities that have teams, such as Providence, Rhode Island, which is close to Boston, and Camden, New Jersey, which is close to New York. I never unpack my suitcase.

The year 1958 ends with a win against the Syracuse Nationals in Philly, stopping a five-game losing streak. I score 24, my average, and wave to the cheering crowd as I jog off the court. We're not setting the world on fire—our record now stands at 12–20—but in the dressing room, Bob Short toasts us and the New Year with champagne. We play fast-paced, exciting basketball, and we're winning some games. But I know we can do much better.

We clink champagne flutes and the owner gushes over our team. For the first time, he sees a future for the Lakers in Minneapolis.

• • •

From day one of training camp, I fit right in with my teammates. I have never been a part of such a welcoming cast of characters. To start with, we have Jim Krebs, our backup center, who walks around under a constant dark cloud, viewing the world as one of doom and gloom. He lugs around a Ouija board, which he believes in deeply and uses religiously on the plane. He repeatedly encourages me to join him in asking the Ouija board pointed questions, such as "Will we advance out of the first round of the playoffs?" or "Will this plane crash?" I refuse. Call me superstitious or a Ouija board nonbeliever, but I don't want any part of the thing.

We play cards a lot, sometimes gin rummy, a little bridge, but mostly poker. I'm not a good card player. I'm an *excellent* card player. I learned by watching my mother and her Sunday afternoon church ladies, every one of them a cutthroat card shark. Those women didn't play for money; they played for blood. And that's how they taught me. When I play with the guys on my team, I'll do anything to win. I rely on certain diversionary tactics I've developed over time, my number-one tactic being that I don't stop talking. This strategy works. Drives the other guys crazy, throws them off. They can't concentrate. I'll take any advantage I can get. I never trash-talk my opponents when I play basketball; I only trash-talk my teammates when I play poker. After a bad beat, claiming he was distracted by my nonstop commentary, Krebs folds a straight against my pair of tens and dubs me "Motormouth." The name sticks.

Another character, Ed Fleming, loves to pull pranks. Ed sucks you in. He seems so innocent and so nice, always smiling and laughing, but don't be fooled: he's planning something. Just hope you're not his next victim.

Sometimes he acts in self-defense. One of the guys on the team—I can't recall who, actually could be a couple of guys—snores like a racehorse. He keeps his roommate up and drives guys nuts on flights. One night, late, around two in the morning, with Ed leading and directing us, a couple of us circle the snorer while he's sound asleep, lift his bed, carry him and the bed into the hallway, and leave him there. The snorer stays in the hallway the entire night. He doesn't wake up until a guy delivering room service careens around a corner, slams into the bed, and drops his tray with a crash, jerking him awake.

Another character—I no longer remember which one—tells us he's bought a dry cleaner's. To support him, a bunch of us give him our uniforms to clean. I think he's overcharging us, but I don't want to complain. One night, before tip-off, he rushes into our locker room late, nervous, and without our uniforms. He lied: he didn't buy a dry cleaner's. He's just been bringing our uniforms to a local dry cleaner and pocketing the difference. This time, though, the dry cleaner doesn't have our uniforms ready. Busted, our teammate pays us back, with interest, absorbing an avalanche of verbal abuse from all of us while we wear our road uniforms at a home game.

As the season progresses, the guys and I become tight. We become more than teammates, more than friends. We form a kind of fraternity. Almost like a family. On the road especially, we do everything together.

Well, almost everything. The first few months, I gravitate toward the other rookies, Boo Ellis and Steve Hamilton, with Ed Fleming often joining us for dinner. Without ever discussing the choices, we avoid certain restaurants that we've heard refuse to serve black people, especially in towns like St. Louis and Cincinnati. When I hear about those places, I have almost a physical reaction.

The memories rush in, tattooed on my soul. I remember the segregated lunch counters in Washington, D.C. I remember the blacks-only playgrounds and parks. I remember the police cruisers that circled our neighborhood, rounding up and hauling away random kids on the street, including my brothers. I want to believe that we're past that, that we're better than that. But as Bill Russell says, if we're making progress, it's slow. I hear it when I play basketball in certain cities, even though I try to drown out the crowd. The racist taunts. The names. The hatred. Warming up in St. Louis with Ed and Boo, already feeling cranky because we've lost a couple in a row, I hear a guy in the crowd shout, "I didn't come here to see the Harlem Globetrotters!" Integration has become law. But how do you legislate against hate?

On the road, after games, the team socializes, getting drinks, going out to eat, and then coming back to the hotel to play cards, either all of us or a few of us, depending on who has "plans." Certain players—namely, Hot Rod, Slick, and Garmaker, whom I call the Three Musketeers—always have "plans." Their lifestyle defies anything I've ever seen. They play as hard at night as they do on the basketball court. Just thinking about how they party exhausts me. Once in a while, Vern Mikkelsen, our captain, the oldest guy on the team, in his early thirties, joins them. When I see him in the morning after he's gone out with the Musketeers, he looks at me through bleary, bloodshot eyes, his face puffy and red, his craggy voice barely audible, muttering that he's too old to run with these guys. He can't keep up.

"They're . . . crazy," he says, grimacing.

I don't know what he means, and I don't want to know. I do know this: Hot Rod, Dick, and Slick pull all-nighters as a matter of habit. I know this because I like to get up early, grab breakfast, and read the morning papers. More than once, I've bumped into the three of them coming into the elevator, finally calling it a night.

Although Hot Rod boasts the most about his partying, Slick may have him beat. One morning, around eleven, the team gathers in the lobby before we head over to the arena for practice. Kundla does a head count and comes up short. He frowns. "Who's missing?"

"Slick," someone says, to stifled laughter.

"Probably still asleep," someone else says.

Probably not in yet, I think. I volunteer to knock on Slick's door and wake him up. Of course, I'm just buying Slick time. I hope that by the time I get to his room and knock on his door, he'll show up in the lobby.

He doesn't.

We leave the hotel and disperse toward several cars to caravan to practice. Out of the corner of my eye, I see Slick running down the street, weaving between parked cars. As he gets closer and I can make out his features, I see that he looks terrible. His hair sticks up in ten different directions, as if he's stuck his finger in an electrical outlet. His clothes, rumpled and wrinkled, look as if he's slept in them. He ducks behind the fins of a parked Cadillac and waits for Kundla to turn his back. The coach squints at the hotel and shakes his head. "The hell with him," he says, and starts heading toward his car.

Slick darts out from behind the Cadillac. He lurches over to Kundla. "You're not looking for me, are you?"

Kundla recoils and waves his hand in front of his face. "Damn it, Slick, you smell like a brewery. Are you just coming in?"

"What? *No.* Been up for hours. I'm coming in from a nice *walk,* because I had such a big breakfast. Pancakes, waffles, eggs, toast, hash browns . . . Did I say eggs . . ."

Kundla glares at Slick. "You're so full of shit. Get in the car."

"You got it, Coach. Yes, sir."

He pulls open the passenger-side door of a nearby car and gets in. The driver, a young woman, screams. Slick rolls out of the car, then looks questioningly at Kundla as she drives off. "She's not with us?"

• • •

On January 16, a cold, icy Friday, after dropping our fourth game in a row, 111–105, to the Syracuse Nationals, we arrive in Charleston, West Virginia, Hot Rod's hometown, to take on the lowly Cincinnati Royals. For some time now, Hot Rod has been lobbying the league to play a game in West Virginia's capital and largest city, and the city's bigwigs —along with some corporate sponsors—have granted Rod his wish. Hot Rod, a legend at Charleston High and a star at the University of

West Virginia, says he's been talking me up to everyone in the city, from his former coaches to business leaders to the mayor.

"I told everyone about you, baby," he says to me. "The whole town's coming. Going to have a packed house tonight."

I'm glad we'll be playing in front of a good crowd, but what's more important is that we need a win. Cincinnati, with a record of 8–32, the doormat of the division, appears to be a team we can handle and thus end our losing streak, even though, after playing five games in six days, we're all dog-tired. Around four in the afternoon, as we drag ourselves into the lobby of the hotel where we'll be staying, I calculate that I'll have barely enough time to grab a quick shower and a short nap before our eight o'clock tip-off.

I splinter off from the team and lean against a wall about ten feet away from Vern Mikkelsen as he walks up to the front desk to check us in. I close my eyes and zone out, paying no attention to the conversation between Vern and the front desk clerk. After thirty seconds or so, I hear what sounds like an argument. I open my eyes and see Vern leaning into the desk clerk, towering over him and shouting. Vern's waving his arms, and his usually snow-white Scandinavian complexion has gone beet red. I push off from the wall, ease in behind Vern, and face the desk clerk.

"Is there a problem?" I say.

Vern speaks to me but has his eyes fixed on the desk clerk. "Yeah. There's a problem. He says I can stay here but you *colored* guys can't."

I feel my lip tremble. I stare at the desk clerk. "Is that what you said?"

"What the *fuck*." Hot Rod springs forward and pushes between Vern and me. He jabs his finger at the desk clerk's eyeball. "You listen to me. You know who this is? This is Elgin Baylor. Now you find us rooms. *All* of us."

The desk clerk, a white guy in his twenties, squirms inside his too-big sport coat with the hotel's emblem stitched over the pocket. "It's policy," he says. "We don't allow—"

"You little nothing," Hot Rod says. "This man makes ten times more today than you will ever make in your life."

He bounces away from the front desk, scans the lobby, spies a pay phone, and charges toward it.

"I'm sorry, Elgin," Vern says.

I'm too incensed to answer. I want to punch my fist through a wall. I want to stomp somebody. I want to scream *"I've had enough!"* But I don't move. I can't move. I look across the lobby at Hot Rod screaming into the phone. He hollers something indecipherable, slams down the receiver, dials another number. At that moment, John Kundla, who's been making transportation arrangements to the arena, comes in, heading toward the front desk. Vern intercepts him and, gesturing wildly, explains the hotel's policy. Behind him, Hot Rod slams down the phone again and joins the rest of us.

"I called two other places," he says. "Same situation. No rooms available."

I glance at Boo Ellis and Ed Fleming, the other black guys on the team. They look shell-shocked. I turn back to Vern, Rod, and Kundla. "You guys can stay here if you want. Boo, Ed, and I will find some other place—"

"No," Kundla says. "We all stay together."

Slick, Krebs, and the rest of the guys huddle around us. They all nod in agreement, every one of them.

"We're a team," someone says.

"I'm going to call Mr. Short," Kundla says.

As he picks up the pay phone, I suddenly feel queasy. "I got to get out of here," I say, eyeing the door. I start to rush outside, but I feel everyone looking at me and I stop. My legs feel so wobbly, I'm not sure they'll carry me to the door. I stand in place, swaying slightly, the lobby starting to blur.

A minute later, Kundla, shaking his head all the way from the pay phone to our huddle, rejoins us. He says Bob Short lost it. Enraged, he told Kundla that he is going to call the hotel manager and demand an apology, and if that gets him nowhere he's going to call the hotel owner. "He might buy the hotel," Kundla says.

But for now, Short insists that we find a place that will accommodate the whole team, and quickly. We have a game to play in three hours.

We find a place. Edna's Tourist Court. A small, clean motel in the black section of the city. Not only does Edna's take black people; Edna's seems to take *only* black people. I'm not sure how we found Edna's, but I'm so angry and tired, I don't care. As we check in at the front desk—more of a front table—Vern Mikkelsen, Slick Leonard, Dick Garmaker, and several of the other white players scope out the people sitting in the lobby and the guests heading toward their rooms. Every person we see is black.

"I don't know," Garmaker says to me.

"What's wrong?"

"Nothing," he says. "I just feel really uncomfortable."

"Now you know how I feel," I say.

When I get my key, I go to my room, shower, and try to nap. But I can't sleep. I stare at the ceiling and replay what's just happened. I feel both heartsick and enraged. I want to disappear. Then I want to shout. I want people to know how outraged, saddened, and sickened this city has made me feel. But what would I say? And to whom? And even if I did say something, who would listen?

Nobody.

Who would listen to a highly paid professional black athlete complaining about being the victim of racism in a small southern city?

Who would care?

I know. I've just got to do my job.

Shut up and play, boy.

Boy.

A couple of hours later, I stare into the ceiling and picture my father and those cops standing over him. He reaches for the leather strap. I see Columbia's face . . . All these thoughts—all these emotions—pummel me as I sit in the dressing room before the game against the Cincinnati Royals.

I look into my locker at my uniform, at my number, at my name. Baylor.

Nothing makes sense. Nothing matters.

I feel so small. I feel so . . .

Dismissed.

And then, as I sit in the dressing room, my body slumped, studying the floor, my anger building, the sour taste of bile rising into my throat, my stomach in a knot, I know exactly what I must do . . . what I will do.

What I will *not* do.

I will not play. I will stage my own private protest. For myself, if nothing else.

I only then become aware of the commotion around me: Locker doors slamming. Voices rising. Music playing. Laughter. Shouting. A swirl of motion. Bodies brushing by me, hands touching me.

I don't respond. I don't raise my head. I sit facing my locker, still wearing my street clothes. Not moving.

Unmoved.

A shadow dusts the wall in front of me.

"Hey."

Hot Rod. Standing next to me.

"What are you doing? You're not dressed."

"I'm not playing, Rod."

I pause, and then, as if I'm speaking the words for the first time, I say, pressing each syllable, "I'm not going to play."

Rod holds a beat and then says quietly, "I made a lot of promises to a lot of people to make this happen. I bragged about you. How good you are. How special. People are coming here to see you. It's awful out. Snow, ice—brutal. People didn't care; they still came. We're sold out. Just to see you."

"I'm not playing, Rod," I say.

Rod swallows. "I thought we were friends." He lowers his voice even more and says, "Elgin, I'd really like it if you would play."

I finally raise my head. "Rod, yes, you are my friend. But I'm a human being. I want to be treated like a human being. I'm not an animal put in a cage and let out for the show. I am not an animal. I am a *human being*."

The world goes cold.

Silence.

I fix my eyes on Rod's. "You don't understand, do you?"

"I'm trying to."

I turn away. I feel a tear slide down my cheek. I flick it away with my thumb. "I know you can't. You can't understand. You don't have black skin. You can't know how I feel."

Through the walls I hear a muffled announcement over the public address system, followed by an eruption of crowd noise. Hot Rod shifts his weight. I know it's time.

I look down at my hands. For the briefest moment, I picture myself in church, my hands folded in front of me in prayer. A wave of absolute certainty streaks through me, followed by a sense of pride, of dignity. I will not waver. I don't know whether I will stay here in the dressing room or join my team on the bench. But I will not play.

Rod inhales and rakes his hand through his hair. He exhales slowly. "Don't," he says.

I search his face.

"Don't play," he says. "You're right. I was wrong. Baby, don't play."

He puts his hand on my shoulder.

I sniff. I nod. We don't speak for a solid minute.

"We're probably gonna lose without you," Rod says finally.

"Against the Royals? Try not to."

"Elgin," Rod says. "I may not understand everything you're feeling, but I understand pain."

"I'm sorry, Rod," I say. "I know you promised a lot of people."

"I did, yeah," Rod says. "Fuck 'em."

Rod squeezes my shoulder and jogs out of the dressing room. He passes Boo Ellis and Ed Fleming, who stand in the doorway. They look scared as hell.

"We don't want to play, either," Boo says.

"It's not right," Ed says. "We're gonna sit out, too."

It hits me then for the first time: sitting out will have consequences.

I have signed a contract, and I could be suspended or fined for refusing to play. I might even be terminated for breach of contract.

I'll take my chances. I believe Bob Short will back me. I know he won't fire me. At least I don't think he will. He's invested too much in me. The team has made a commitment to me. Some people call me "the franchise." I'll be all right. But I worry about Boo and Ed.

"Obviously, I know how you guys feel," I say. "But you should play. If you don't, you could jeopardize your careers."

They take this in. They stand in the doorway, considering what I've said.

"You shouldn't be in this position," I say. "None of us should. But I don't want you to suffer because of me."

"I don't know what to do," Boo says.

"Play," I say.

"I think we have to, Boo," Ed says. "We don't have a choice."

I shoot them a grim smile. "Do you guys have any idea how pissed off I am right now?"

Boo and Ed decide to play; they get into the game for a few minutes each. Even though the Royals are probably the worst team in the league, they beat us 95–91, sending us to our fifth defeat in a row. Wearing my street clothes, I watch from the bench. During the game, rumors fly. *Baylor's out because he suffered a last-second freak injury. Baylor has the flu. Baylor came down with food poisoning. Baylor broke a team rule and has been suspended for the game.*

It's not until after the game that everyone learns that I chose to sit out in protest for being denied a room at a hotel in Charleston because I was black. Then, for forty-eight hours, all hell breaks loose.

The local press swarms all over the story, then the national media piles on. Charleston mayor John Copenhaver tells *The New York Times* that he regrets the incident but refuses to apologize. He says, "The incident was something over which our city has no control."

The American Business Club of Charleston files a complaint with the NBA and with league president Maurice Podoloff that "urges disciplinary action against Elgin Baylor." This organization claims that my absence from the game embarrassed the city and damaged Charleston's chances of hosting future NBA games. Someone from this group also says that I cost the city money, potentially as much as $800 in ticket sales. That's completely untrue, since nobody knew I'd decided to sit out the game until it was over. Other people call my actions "inexcusable," and a city official from another town says that I should apologize to the city of Charleston.

Then I receive letters. Mostly hate mail. Threats. Many come from virtual illiterates. Most express the same thought in different words: "Nigger, who do you think you are?" I toss these into the trash.

I do open a few positive letters. Several black people thank me for standing up to prejudice and injustice. Several white people applaud me for my courage.

Two days after the game, league president Maurice Podoloff says the NBA will take no action against me, stating that the "incident is between the player and the club owner."

The club owner, Bob Short, supports me publicly, privately, and in the press. He says that I will not be disciplined in any way, allowing that the incident in Charleston caused me "great emotional stress."

Then, a couple of days afterward, to my surprise, Mayor Copenhaver calls me and offers a personal apology. I guess he had a change of heart, or an attack of conscience, or maybe some prominent black businessmen got to him. In any case, I thank him for his apology.

I'm thrown by all the attention. I didn't intend to cause a firestorm. I wasn't trying to start a revolution. I don't see myself as political or controversial. But my action has an impact. Because of my protest, the league creates a new policy, forbidding hotels to discriminate against any member of an NBA team. Two years later, in 1961, I will play in an All-Star Game in Charleston. I will also accept an invitation to stay in the same hotel that turned me away, this time receiving nonstop apologies as I check in.

I'm glad what I did changes some things, yet I don't see what I did as anything but standing up for myself. That's all.

It comes down to this: I want to be treated the same as everybody else. I will not allow anyone to demean me, diminish me, or belittle me. I'm a man, same as any other man — no more, but no *less*.

• • •

After Charleston, we fly back to Minneapolis for a game against Philadelphia. I don't talk much on the flight or during the game. I just play. I score 30 and we win 119–98, finally ending our losing streak. On January 23, the sour taste of Charleston still lingering, I head to Detroit with my teammates Dick Garmaker and Larry Foust to participate in the NBA's annual All-Star Game. I sit on the bench before the game starts, my eyes involuntarily wandering over the crowd, scanning the ten thousand fans jammed into the Detroit Olympia stadium, wondering how they will react when the announcer says my name. I hear my name, and the crowd noise builds to cheers and applause. My mind still reeling, my emotions still raw, I jog out to join the other members of the West team. Slightly on edge, a rookie invited to play with veterans and perennial All-Stars, the best of the best, I will myself to block out any boos smattered among the cheers and respond the only way I know how: by putting my head down and playing. I score 24 points, our team beats the East 124–108, and I'm named co-MVP with Bob Pettit.

After the All-Star Game, the Lakers alternate winning and losing until we get hot and reel off five straight victories. We cruise through February, including, on February 25, a 116–96 drubbing of the Cincinnati Royals, a game in which I score 55. We end the season with a 33–39 record, a fourteen-win improvement over the previous year. Vern and some of the other players seem elated. Unlike last year, we're heading into the playoffs.

"The town's hungry for a winner," Vern tells a reporter.

I'm even hungrier. I vow to take the team deep into the playoffs.

We open the playoffs at home with a best-of-three series against the Detroit Pistons. The Pistons play hard-nosed, gritty basketball. I know a couple of their players from D.C. I played against their best player, Gene

Shue, in a tournament when he was in college and I was in high school. My Stonewalls torched his team, and I outscored Gene, but Gene can play, and he has proved it in the NBA.

I also know their sixth man, Earl Lloyd, from the playgrounds. Earl, several years older than me, holds the distinction of being the first black player in the NBA. I like Earl as a person, and we've become good friends, but, man, I hate playing against him. Especially when he checks me.

Six foot six and 250 pounds of sheer muscle, Earl plays rough and gets into a lot of altercations. I'm sure he's taken a lot of crap for being the first black player in the league. I know he's heard ugly things from fans and players and lashed out, literally. I also know that Earl likes to fight. I played with some very tough guys on those D.C. playgrounds, but nobody would mess with Earl. We nicknamed him "the Policeman" because he'll harass you first and ask questions later.

In the first quarter of game one, Earl comes off the bench to guard me. I know the coach has told Earl to do whatever it takes to stop me, and he does. He bumps me, shoves me, clubs me, hacks me, elbows me, pounds me, and practically tackles me.

"Damn, Lloyd," I say one time, after he whacks me across the arm. "If you weren't my friend, I'd punch you in the mouth."

"What? I'm just playing you *tight*."

"You're mugging me. You want my wallet? Let me go get my wallet."

Earl looks at me as if I've hurt his feelings. "Come on, man, you know I'm not trying to hurt you."

I know that's true. Earl just enjoys physical contact.

We win a close, rugged war by three points. I score 14, and after the game I soak in a long, hot bath, every joint, every muscle, every square inch of me aching.

We lose the next game in Detroit, then come back to Minneapolis for the winner-take-all final game. I shake loose for 30 and we batter the Pistons 129–102, winning the series.

I hug the Policeman after the game and join my teammates in the locker room for a celebration. Nobody's popping champagne yet, but we've done something Lakers fans haven't seen in some time. We've made the playoffs and advanced.

• • •

The St. Louis Hawks, led by their dominant front line of First Team All-NBA selection and perennial All-Star Bob Pettit, All-Star Cliff Hagan, and muscular center Clyde Lovellette, hold the Western Division's best record. In our twelve regular season matchups, we've managed to beat them only four times.

That's the regular season. For me, the season starts *now*.

We go into St. Louis for game one of a best-of-seven series, and, in front of a packed arena of screaming, cheering, and jeering fans, we lose by 34 points.

The loss fires me up. "It's only one game," I say in the locker room to my dejected teammates. "Shake it off. We'll get 'em at home."

We do.

Game two. Taking a page from the Policeman, I fasten myself to Bob Pettit. He can't shake me. Several times, he glares at me as if he wants to come at me. I ignore him, say nothing, and stick to him. We win 106–98. I score 33 and frustrate Pettit, holding him to 22, well below his average.

We return to St. Louis for game three and it's as if the city, the fans, and the team have put some kind of hex on us. We can't buy a basket. The hoop looks about the size of a paper cup. When we play defense —well, we really don't play defense. It's like the Hawks are out there by themselves, running layup drills. They demolish us 127–97, in front of a belligerent crowd, and take a 2–1 series lead. In the two games we've played in St. Louis, the Hawks have buried us by a total of 64 points.

At home, playing in the cavernous Minneapolis Armory, in front of more fans than we've had all year, we become a different team. We pull ourselves together. I check Pettit again, this time limiting him to 21 points. I score 32, and we beat them 108–98 to knot the series, 2–2.

In the dressing room afterward, slowly slipping on my street clothes, I mumble the obvious, a thought that's become my mantra: "We have to win a game in St. Louis. Somehow."

March 28. Downtown St. Louis. I pick at an early lunch, scan the morning papers, linger in the hotel lobby watching people, and then go up to my room for my usual pregame nap.

I can't sleep. I blink at the ceiling, replay our last two blowouts here, get out of bed, and pace. I wish we could play right now. I know we will beat the Hawks. I know it. We have to. I can't explain it, but I feel as if my young career will be defined by this game.

In the St. Louis arena, the din of the hysterical crowd making the floor planks vibrate, Krebs and Lovellette jump center, the Hawks win the tip, and the game begins. We exchange leads through the first quarter. The game takes on a life of its own. It feels as brutal and tense as a heavyweight fight. After being shut down in the previous two games, Pettit seems to have found a new gear. He sprints behind picks, shooting jumpers on the run, and drives toward the hoop as if his life depends on every shot. I glue myself to him, but he's unconscious. In basketball, good offense beats good defense every time. I can't stop Pettit, and he, Hagan, and the other Hawks can't stop me. I blitz the Hawks, leading fast breaks, draining long jump shots, and showcasing a variety of moves that leave defenders in the dust. On one play, I whirl, I fake, the Hawk defender flies by me, and I bank in a fifteen-footer.

We go nip and tuck into halftime, stay close at the end of three quarters, and by the end of regulation—both centers, Krebs and Lovellette, having fouled out—we've battled to a 90–90 tie.

My legs feel so tired they tremble. I sit on the bench and drape towels over my quivering thighs. I take slow sips of water, don't swallow, spit the water out. My chest heaves. The buzzer sounds. I stand and begin the long walk to the center circle.

Don't let this one slip away.

I draw first blood in overtime, sinking a short, slashing step-away jumper. We cling to this advantage by what feels like bloody fingernails, finally eking out a 98–97 win. I match Pettit point for point, each of us scoring 36. The game ends and I dash into the locker room, slap the door frame, and muffle a shout. We haven't actually won anything yet — we lead the series 3–2 and have to win one more game to advance — but in my mind we've already won. By beating the Hawks at home, we've cut out their hearts. We will finish them off in Minneapolis. I know it won't be easy, but we have them now — we *have them*.

We return home and play game six the next night, giving us no time to catch our breath. This game feels like a replay of game five, a back-alley brawl. In the end, we win by one basket, 106–104, shocking the Hawks, stunning our fans, and advancing to the NBA Finals against Bill Russell and the Boston Celtics.

• • •

April 4, Boston Garden.

We take the court in front of fourteen thousand insane and probably inebriated fans. This crowd roars so loudly, I swear the rickety building sways.

Before the game, in our icebox of a locker room, I don't give the guys a pep talk — that's not my style — but anyone could read my body language.

I think we can beat these guys, my eyes read. *We can take them.*

I have no logical reason to feel this way, since we haven't beaten the Celtics all season, losing all nine games we played against them. The game that I can't shake, the craziest game — maybe the worst loss of the season — came on February 27 in Boston, two nights after I dropped 55 on the Royals and we ended our five-game losing streak. We put up 139 points against the Celtics — I had 28 — and we still lost by 34. The Celtics hammered us 173–139, setting a record for the highest point total ever in a regulation game. Not a shred of defense spotted anywhere, mainly because Bill Russell was nursing an injury and didn't play. And our defense? What defense? Seemed like every player on their team scored 30.

Still, I have high hopes. The Celtics have just come off a hard-fought

seven-game series with Syracuse, and I believe they're tired and vulnerable.

We go at them, hard. At the end of three quarters, we've locked them
up and tied the score, 84–84, after trailing for much of the first half.
Finally, though, the Celtics pull ahead and squeak out a 118–115 win.
I score 34, but they simply have too many weapons, too many shooters
and ball handlers, and of course the force in the middle, Bill Russell,
who snatches every rebound with a fury, blocks a dozen shots — tapping most of them to teammates — and triggers countless fast breaks
by snapping off outlet passes to Cousy or Sharman, often at half-court.
Russ scores only five points himself, but six other Celtics score in double figures, led by Frank Ramsey with 29 and Tommy Heinsohn with
24. That's how the Celtics dismantle you. You may stop one or two of
them, but four others will come at you, and they always have Russell
protecting the rim.

We drag ourselves into the dressing room the next night, feeling
wasted and dejected. I don't see how we can play any better against the
Celtics. We can't match Russell in the middle, and our guards can't run
with their guards. We just don't have any answers. Kundla, forced to
say some words of encouragement before the game, mutters something
about "We're only down one, and it's a seven-game series," but nobody's
buying his halfhearted rally speech.

As game two starts, I see they've switched up their defense from the
previous game. Apparently they've decided to concentrate on me. They
come at me with a rotation of players, starting with Heinsohn, switching to Ramsey, then Jungle Jim Loscutoff off the bench, and occasionally assigning their slick six-four guard Sam Jones to chase me. A couple
of times they even try a kind of zone trap with two players: one in front
of me, one behind me. You're not allowed to play a zone defense in
the NBA, but I glance over at their coach, Red Auerbach, and I catch a
glimpse of his self-satisfied smirk while he snips off the tip of his victory
cigar midway through the fourth quarter.

His rotating, smothering defense works. I manage only 13 points, and Boston blitzes us, 128–108, as six Celtics again score in double figures. Russell, refusing, as usual, to acknowledge me during the game, scores nine points, all that he needs.

Down 2–0, losers of eleven in a row to the Celtics, we travel back to Minneapolis for game three. I feel completely frustrated. I know I shouldn't. Nobody expected us to make it this far. We didn't even play .500 ball. We had no right to beat St. Louis and get into the Finals. I try to convince myself of all of this, but it doesn't help. I want to beat these guys. I love Bill—he's my friend—but, man, I want to whip his ass.

In game three, I give it my all, but I play like garbage. The Celtics employ that same defense by committee and I score only 14 points in a 123–110 rout that's not as close as the score. Russell scores only eight points, but three other Celtics score over 20, and by the middle of the fourth quarter Auerbach is leaning back on the Celtics' bench, his hands locked behind his head, puffing away on his fat victory cigar. I feel like dashing over there and swatting that stinking stogie right out of his arrogant mouth.

I know I probably should consider the defense the Celtics play against me a compliment. Auerbach is making a statement, telling me, his team, and the entire basketball world, "I'm not going to let Elgin Baylor beat us."

Well, I've had it. We may not have enough pure talent to take out the Celtics this year, but they're not winning the NBA championship without a fight. They've frustrated me twice. They will not stop me tonight. I want to win at least one game against them, and I want to do it in front of our fans.

I arrive at the Armory more focused and determined than I have for any game all year. I pull my car into the parking lot at the same time as Slick and a couple of the other guys. As I get out of my car, I see that Slick and the others have packed their cars with clothes and boxes, as if they're moving somewhere.

"What's all this?" I say.

Slick shrugs. "Let's face it, Elgin. We're down three games to none. We probably won't win tonight. We figured we might as well head home for the summer after the game."

Before I can speak, Slick says quickly, "We just want to get a head start. Don't worry, I'm going to play hard, like I always do. I'm just a practical guy."

Slick does give it his all. He scores 23 points. I ignore any defense the Celtics throw at me. I want to win this game so bad, I play as if I'm playing for my life. I slip around people, push through people, fake my way around people, fly by people, jump over people. I actually hear some *ooh*s and *ahh*s from the disappointing turnout in the stands. I score 36, but we lose 118–113, and Boston has its second championship in three years. Red, maybe out of the slightest flicker of respect, waits to light up his victory cigar until right before the final horn sounds.

I find Bill after the game. We hug, and he whispers, "You played your heart out. We just had too much, man."

"I know. Thanks."

"So," Bill says, nodding his head solemnly, trying to fight the emergence of his killer grin, "where we gonna eat?"

. . .

I end my first season in the NBA averaging 24.9 points, 15 rebounds, and 4.1 assists per game. I win the NBA Rookie of the Year award, finish third in the MVP voting, and am named to the All-NBA First Team, along with Bob Pettit and three Celtics — Bob Cousy, Bill Sharman, and Bill Russell — marking the first time two black players have made the All-NBA First Team. To show his appreciation, Bob Short doubles my salary to $50,000, raising me into the same elite category as Russell, Cousy, Pettit, and a few other top players in the league.

Then he gets to work on the new season.

First, he fires John Kundla. I can't say I'm shocked. I like John, but at some point during the season he seemed to become more low-key than usual and even a little . . . lost.

I *am* shocked when Mr. Short calls to ask my opinion of the guy he wants to hire to replace Kundla.

"What do you think of John Castellani?" Short asks me.

Castellani. My college coach. Hot-wired. Super high-energy. A little rigid. Not that experienced. I picture him calling me Edge. I smile. "I like John a lot."

"Do you think he can coach this team?"

"Well, the NBA and college . . ." I say, letting my voice fade. "They're much different. And he's pretty young."

"Agreed. How do you think he'll do?"

"I honestly don't know," I say.

Soon after our conversation, Bob Short coaxes John Castellani out of "basketball retirement" and hires him as the coach of the Lakers.

Then they work on improving our roster. Vern Mikkelsen retires, but we keep our core group of Foust, Hot Rod, Slick, Krebs, Boo Ellis, Ed Fleming, and Steve Hamilton. We sign shooting guard Frank Selvy and add two top prospects in the draft: Rudy LaRusso, a six-foot-seven power forward from Dartmouth, who will become a good friend, and Tommy Hawkins, a six-foot-five forward from Notre Dame, who will become my closest friend on the team.

Then, out of the blue, Bob Short encourages me to enlist in the army.

• • •

"You're going to be drafted," Mr. Short says. "Just like Elvis. Could be any time now."

When Bob Short talks about the military, I listen. Mr. Short, a former navy commander, served as undersecretary of the navy. "The army's looking to draft celebrities and big-name professional athletes," he says.

I haven't given much thought to serving in the military. Like everyone else, I followed Elvis Presley's induction into the army. When he got drafted, a year ago, the army offered him a two-year special assignment, telling him he could live stateside in spruced-up, higher-end housing and not in a barracks, and for his "duty" he would only have to perform occasionally for the troops. Elvis refused any entitled treatment, saying he wanted to be treated like any other soldier. The army complied and deployed Private Presley to an army base in West Germany, where he was now serving out his two-year military commitment.

"What do I do?" I ask Mr. Short.

"It would be unpatriotic not to serve your country," he says.

"How can I play basketball if I'm sent overseas for two years?"

"I have a solution," Short says. "I think you should sign up for the Army Reserves."

After Bob lays out his idea, I talk over my options and my military obligation with Ruby. We've both flourished in Minneapolis. She's completed her junior year at the University of Minnesota and has settled in with a close circle of friends. We've just found out that we're about to start a family. Signing up for the Reserves, as Bob Short explains, will allow me to serve my country and play basketball at the same time. In these days of peacetime, I would become a weekend warrior, with six years of summer assignments. I would be required to complete six weeks of basic training, and do so immediately, and eventually six months of active duty at a base somewhere in the United States.

"When would that be?" Ruby asks.

"Not for a while. They won't tell me exactly when."

Finally, we agree to Bob Short's plan. Rather than wait to be drafted and sent overseas for two years, like Elvis, I will join up now for a longer but less intensive commitment.

But even as I sit at the desk in the recruiting office and fill out the paperwork, my head swims with doubt. I may be somewhat known in basketball circles, but I'm no Elvis Presley, and I certainly don't seem like an advertisement for the military. My face will never appear on a recruiting poster. I really shouldn't have been admitted into the army at all.

I'm married, with a child on the way. At six foot five, I fall one inch short of the maximum height limit. And I'm color-blind. I can't differentiate shades of green, or all the various browns, or identify khaki. All of those colors look the same to me. If we ever got into a firefight, I'd probably shoot my own guys.

Won't the army notice all this, kick me out of basic training, and send me home? No. Uncle Sam wants me.

The army dispatches me to Fort Sam Houston, outside San Antonio, for my six weeks of basic training. Talk about a turnaround: I've gone from winter in Minnesota, where it's as cold and icy as a glacier, to summer in Texas, where it feels as if I've landed on the surface of the sun. But Bob Short makes good on his promise. As we prepare for the start of our 1959–60 season, he moves the Lakers' summer training camp to San Antonio so I can work out with the team on weekends and afternoons when I get permission.

I don't want to call my six weeks of basic training a joke, but, okay, it's a joke. I feel like I'm an extra in a movie Columbia took me to, where Abbott and Costello join the army.

On our very first day, the new recruits wake up to the sound of a horn blaring in our ears at 5:00 a.m. We get dressed, make our beds, form a line, and stand at attention for inspection. The door whips open and our drill sergeant clomps into the barracks. As he slowly approaches, I do everything I can not to stand out. This is difficult, because I'm wearing my street clothes. I don't have a uniform—nobody could find one that fit me.

I stare straight ahead, hoping that the sergeant won't notice and will walk by me. Then I hear him skid to a stop in front of me. I feel his eyes on me. The sergeant is wiry, bald, and, as I will learn, always in a bad mood. He cranes his neck up at me and shouts, "What the hell is *this?* Where is your uniform?"

"I don't have one, sir."

"Don't have a *uniform?* What do you mean?"

"Never got one, sir."

"You missed receiving your requisition? Where were you? Were you too late?"

"No, sir, I was too tall."

"What?"

"They didn't have my size, sir. They're working on it. They're sending out for one or something. I don't know exactly what they're doing, sir."

He gets right up into my chest. The top of his head brushes my chin.

"I know who you are, Private *Baylor*." He spits out the word like it's an olive pit and then he pauses. "Do you know who I am?"

I hesitate. "Yes—"

"No, I don't think you do. I'll tell you who I am." He pauses again. "I am YOUR WORST NIGHTMARE!"

"Yes, sir."

"Do you know where I live?"

"No, sir."

"I live UP YOUR ASS!"

I peer down at the top of the sergeant's head and try not to crack up. I snort and swallow a laugh.

"Are you . . . *laughing?*"

"No, sir. I have a condition."

The drill sergeant steps back, sees my bed, and stares. Actually, I don't have a bed—I have two beds. The standard army cot can only accommodate someone six foot two or shorter. In order to lie down without my feet dangling over the edge of the bed and hitting the floor, I have slid two cots together. The sergeant continues to stare. Finally, he speaks. "What the hell is *this?*"

"My bed, sir."

"It's *two* beds."

"Yes, sir. I can't fit in one bed."

"You're only *allowed* one bed. One bed to a man. One *bed*." He cocks his head and snarls. "Oh, I get it, Private Baylor. Private *Elgin Baylor*. You think you're special, don't you?"

"No, sir. I was just trying to fit into a bed so I could sleep. I wrote down that I was a basketball player, but they must've thought I was a guard—"

"I'll tell you who you are. You're just another lowly grunt. Exactly like everybody else in this army. Do you understand?"

"Yes, sir."

"I will expect to see you here tomorrow morning for inspection at oh five hundred, IN UNIFORM."

"Yes, sir."

"If you don't have a uniform, Private Baylor?"

"Yes, sir?"

"THEN I DON'T WANT TO SEE YOU."

He holds his scowl for a count of five, turns on his heel as if he's leading a parade, and continues walking down the line.

I don't receive a uniform that day, or the next day, or any day that week. I don't receive a uniform until the day before my six weeks of basic training ends. I do follow the sergeant's order: I stay away from inspection and try my best to avoid him. Having no choice, I wear civilian clothes to every required activity. One morning, I return from completing the obstacle course and find that someone has welded a second bed to my original bed, forming one ridiculously long twelve-and-a-half-foot bed. When the sergeant sees this, he goes ballistic.

"What the hell is *this?*"

"It's my bed, sir. Someone welded two beds together—"

"It doesn't fit in the barracks. It's an eyesore. Get it out of here!"

A couple of guys and I carry my bed into a room set off from the main area of the barracks, and this is where I sleep for the rest of basic training, apart from everyone else. I don't mind. I'm not exactly becoming best friends with my new army buddies.

Although I meet a few nice guys, I feel like an outcast. I find I don't have much in common with most of my recruiting class. At twenty-four, I'm older than almost everyone, many of whom have just graduated high school, some of whom have dropped out. I joined the army to fulfill my military obligation and serve my country. To be honest, I can't wait to get out. A lot of the guys I meet never want to get out. They are gung-ho military dudes and want to make the army their ca-

reer. A few guys come from cities, and I can relate to them. But many
come from the country—the Deep South, the rural Midwest—and
look at me with curiosity, at best. I'm sure they've seen black people
before, but I doubt they've interacted much with any. By unspoken mutual consent, we keep our distance.

I discover a few guys who fall into a category I call "borderline."
These guys like to fight, and they *love* guns. I go out of my way to avoid
them. Sometimes I can't. One guy, a certified psycho, flips out one day
at the shooting range. Lying on his belly, blasting away with his rifle,
he suddenly jumps to his feet, turns his back on the target, and starts
shooting wildly in every direction—into the air, off to the sides, and at
us. I hit the ground, duck and roll as bullets whiz by me. Three officers
converge on the psycho, grab his weapon, tackle him, and wrestle him
to the ground. I never would hear what happened to him. I guess he got
arrested or committed to a mental hospital. Or maybe he became a drill
sergeant.

The long, sweltering, boring days drag by. I count the hours until
basic training ends and I can return to playing basketball full-time, instead of practicing on weekends with the team or shooting by myself for
thirty minutes or so during unbearably humid evenings, on the dilapi-
dated hoop hanging from a pole outside the barracks. I've never loved
to practice, but after sitting alone in my street clothes, twisted like a
pretzel on my absurd welded-together bed, ticking off one more day of
boring-as-hell basic training, I would practice for twelve straight hours
a day before I'd ever go through basic training again.

Mercifully, the six weeks of basic training ends. I have finally received my uniform. I have been promised a proper bed when I serve my
active duty. I have learned how to shoot a gun, though poorly. And the
army has given me an assignment, a job. For reasons I will never understand, I will be a medic. To train for this, I spend hours each day sitting
in musky classrooms without air-conditioning, sweatboxes that cause
me to doze off as the instructor drones on, lecturing about anatomy.
Later, with another instructor hovering over me, I practice first aid on
plastic or foam dummies. I fumble through making a splint and wrap-

ping bandages onto broken plastic limbs. We repeat the routine the next day: back to the classroom, where I snooze through tedious lectures and mind-numbingly monotonous medical training films. After I complete my training as a medic, of which I will recall almost nothing, I pray that I never experience real combat and that I'm never allowed anywhere near an actual wounded human soldier.

．．．

October 18, 1959. We begin the 1959–60 season at home against the Detroit Pistons. I don't know whether I'm jumping out of my skin to play basketball after basic training or I desperately want to eliminate the still sour taste of losing the Finals to the Celtics in four straight, but I go off. I score 52 points against Gene Shue, Earl Lloyd, and newcomer Bailey Howell. I can't miss, but we lose by one measly point, spoiling John Castellani's coaching debut. We win the next game on the road in St. Louis, but then we go on a five-game losing streak, including a terrible double-overtime loss to the Pistons. We beat the Pistons twenty-four hours later and then, on November 8, we go back home to face the undefeated Celtics for the first time since the Finals. We haven't been playing well—our record stands at 2–6—but I wipe that away. I step onto the court against the Celtics feeling beyond determined. We're going to beat Boston tonight. I don't care what it takes.

But I'm not only going up against Bill Russell, Bob Cousy, Bill Sharman, Tom Heinsohn, Sam Jones, and the rest of the formidable Celtics cast. I'm going up against history. The Lakers have lost twenty-two straight games to the Celtics. Admittedly, nine of the losses came before my time, but still, our team hasn't beaten the Celtics in *two years*.

That stops tonight.

It won't be easy.

Red Auerbach goes into his swarming, rotating Stop Elgin Baylor defense, assigning four different players to guard me. As Heinsohn switches off me and Sam Jones takes me, I actually laugh.

You better come up with a better defense, I think.

The more players Red rotates on me, the more I score. I don't know whether John says anything or it just happens naturally, but my teammates keep feeding me. We take an early lead and keep running up the score. After three quarters, we're beating Boston by 30 points. I know we have a quarter to go and anything can happen in basketball, but I wish I had a victory cigar in my pocket right now that I could light up and blow the smoke Red Auerbach's way.

At some point late in the final quarter, with the game well in hand—we're up by 20—John calls a timeout. As we huddle around the coach, Hot Rod whispers something in John's ear.

"What did you say, Rod?" I ask.

"Nothing."

I give him a look. "Tell me."

"Okay, fine. You're one point away from Fulks."

"Who?"

"'Jumping Joe' Fulks. You know what I'm talking about."

I do. Jumping Joe Fulks set the NBA record for most points in a game: 63. He established the record in 1949, ten years ago. The NBA wasn't even called the NBA back then.

"It would be sweet, Edge," John says. "We finally beat the Celtics after two years and you set the new scoring record. Not a bad day's work."

I shift my weight and look past the guys, into the stands. I honestly don't care about the scoring record. But beating the Celtics? That I care about.

"I don't mind sitting out, as long as we stay ahead of them," I say. "Happy to let someone else play. Give a rookie some minutes."

"Yeah, you probably couldn't catch Fulks anyway," Hot Rod says. "You need two whole points."

The team laughs. Even I chuckle.

"Well, let's see what happens," John says. "Stay out there, Edge."

I don't go out of my way to score. I'm content to rebound and set up my teammates. But in the last minute, someone on our team steals the ball and dishes it to Hot Rod, who hits me with a pass. I break for the hoop for a wide-open layup and—

One of the Celtics appears out of nowhere and fouls me. Hard.

I realize then that the Celtics know I'm within two points of the scoring record and they don't want me to break it against them, especially on a layup. If I'm going to get the record, I'll have to do it by making two free throws. They also know that I've been struggling at the free throw line all game. I've missed five of them, which is unusual for me.

I glare at Red. He turns his head away.

I step to the free throw line and take a deep breath. I dribble once, twice, breathe again, bend my knees, and shoot. The first free throw rattles around the rim . . . and drops in. I've tied Joe Fulks. I've scored 63.

I swish the second free throw. I now own the NBA single-game scoring record, 64 points.

The crowd erupts. I acknowledge them with a wave. I turn and glance at Red again. This time he's staring right at me. After a moment, he nods.

. . .

After finally beating the Celtics, we drop the next two games and then head to Philadelphia to take on Wilt Chamberlain and the Warriors. While we've limped to a 3–8 record, the Warriors have started fast, winning eight of their first nine games. Wilt, who will shortly start scoring points at an all-time-record pace, has begun his NBA career a little uncertainly, content to score when necessary and pass off to his teammates.

As the game starts, I come onto the court and shake hands with Wilt. After leaving the University of Kansas a year and a half ago, Wilt played a season with the Harlem Globetrotters. I haven't seen him since my wedding and haven't played against him since we went up against each other on the D.C. playgrounds.

"Been a while," he says.

"It has."

"You look good."

"You look the same. Still pretty tall."

He glances at the other four starters on our team and sniffs. "This time I *will* light you up."

He doesn't light us up, but we lose a close game. Wilt scores 28 and I score 27. That night, we fly back to Minneapolis for a rematch against the Warriors, playing in St. Paul. This time we beat them. After the game, Wilt retreats to his hotel room. I meet Ruby outside the dressing room. She has attended the game, as she does most home games, sitting with the other wives. As we head to the exit, Ruby nods toward the visitors' dressing room. "I was going to invite him to dinner," she says.

"Who?"

"The tall guy."

"I don't think he would come," I say. "He keeps to himself. He's a loner."

We scuffle through the rest of November and into December, managing only three wins in eleven games. Boston comes to the Armory again, now bringing an 18–4 record to our miserable 7–17 mark. This time I score 29, but seven Celtics score in double figures and they rout us 121–104, leaving us smarting. As the season progresses, we will play six more games with Boston and lose every one.

· · ·

I start to hear grumblings about John. Several players speak to me privately. They don't know what to make of him. A few like him and admire his enthusiasm, but they all say that John seems unsuited for the pro game. They point out, correctly, that he needs to realize we're playing a seventy-five-game season, a marathon, not a thirty-game college season sprint.

Pro players respond to a coach who seems steady, cool, unwavering. John appears volatile. Sometimes he seems rigid and insistent, absolutely married to his prepared game plan. But more often he seems unsure. Rudy LaRusso, a Dartmouth grad, tells me that John once asked him if he knew any good plays. That didn't exactly fill Rudy with confidence. Also, John's substitution pattern during games feels scattered. I ask him

At Seattle U, Dick Stricklin and I giving coach John Castellani a "lift" outside the team bus.
© *Seattle University Archives*

I snatch a rebound in the somewhat modest gym of the College of Idaho.
© *College of Idaho Archives*

High flying against the University of Portland Pilots.
Bettmann / Getty Images

Mom and Dad, one of the few times they both saw me play in college.
AP Photo / Marty Lederhandler

At Seattle, looking sharp, on my way to class with a couple of my buddies.
AP Photo / Ed Johnson

Greeting our wonderful, wild fans in Seattle after our NCAA Finals loss to Kentucky. Note my snazzy hat. Seattle Times, *Vic Condiotty*

On the rise, about to bank in a
reverse layup against the Knicks.
Marvin E. Newman, Sports Illustrated /
Getty Images

Now a Los Angeles
Laker, after a game on
the road.
Bettmann / Getty Images

Rising up for a jump shot between two Knicks at Madison Square Garden.
Bettmann / Getty Images

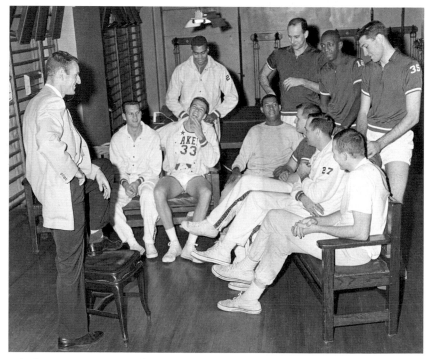

A locker room meeting with coach Jim Pollard and my Minneapolis Lakers teammates. *AP Photo*

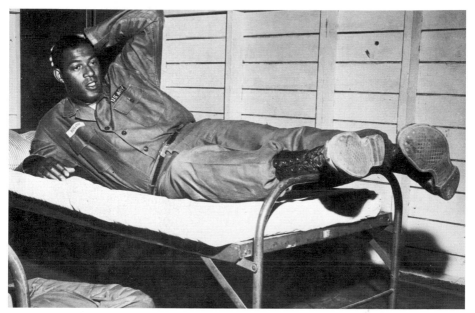

"I'm in the army now." Private Baylor in the barracks, lounging on my too-short bunk.
Bettmann / Getty Images

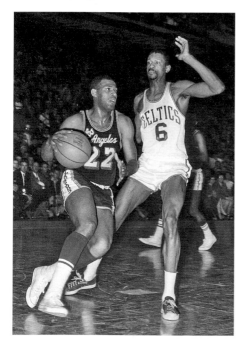

Driving to the hoop against the Celtics, guarded by Bill Russell.

Dick Raphael / Getty Images

Double-teamed on my way for a deuce.

Phil Bath, Sports Illustrated */ Getty Images*

Putting up a baby hook against the Knicks.

Wen Roberts, NBA Classic / Getty Images

Pulling up for a midrange jump shot.

Phil Bath, Sports Illustrated / *Getty Images*

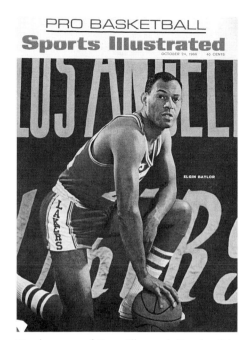

On the cover of *Sports Illustrated*, October 24, 1966, after returning from knee surgery.
Fred Kaplan, Sports Illustrated Classic/*Getty Images*

Swinging a one-hand pass against the Knicks.
Wen Roberts, NBA Classic/Getty Images

At Magic Johnson's retirement, standing with Lakers legends Jerry West, Wilt Chamberlain, Magic, and Kareem Abdul-Jabbar. *John W. McDonough,* Sports Illustrated Classic/*Getty Images*

With my sister Gladys, ninety-one, my only living sibling, at her home in Harpers Ferry. When she saw me play in high school, she called me a show-off. *© Elaine Baylor*

At the Lakers' celebration of my eightieth birthday at Staples Center, with the two loves of my life—Elaine and my daughter Krystle. *AP Photo / Kevin Reece*

about this and he says he'll always stick with the hot guy, but he's still looking for combinations that work. This uncertainty backfires. Guys sit on the bench and squirm, wondering how much they'll play, or if they'll play at all. Once they get into a game, I see them glancing nervously at the bench after they make a mistake, concerned that John will yank them and sit them for the rest of the game.

I feel for John—and my teammates—and I hate that I'm caught in the middle. I advise my teammates to talk to John directly if they have a problem, and when John comes to me for pointers about how to talk to certain players, I tell him that he needs to speak to them directly. I don't want to be a go-between. I don't want to say the wrong thing to anyone or have anyone misinterpret something I say. I also don't want to lose any friendships on the team or jeopardize my friendship with John.

Bob Short doesn't hide his concerns, either. By early December we can tell that attendance at our home games has fallen dramatically. We just don't draw. Between our losing record and the brutal Minnesota weather, sometimes fewer than a thousand people show up. I start to hear all sorts of rumors, ranging from Bob wanting to move the team to a different city that better appreciates basketball to Bob secretly looking for someone to buy the team because, despite all his investments and supposed business savvy, he's going broke.

In either scenario, Bob appears to be frustrated with John. In mid-December, we face a stretch in which we play four games against Cincinnati and Detroit, the two worst teams in the league. Bob gives John what sounds like an ultimatum. We need to beat these teams—especially at home—or certain people may be looking for work.

We do lose to Cincinnati at home, but then we beat the Royals on their court and win both games against Detroit, taking three out of four of Short's ultimatum games. Bob seems mildly placated. But then we go on the road at the end of the month and fall apart. We lose a tight game in Boston, get blown out in Philadelphia as Wilt drops 45 on us, and lose to the Knicks in New York. We come home to play the Syracuse Nationals on New Year's Day, January 1, 1960, dragging with us a pitiful 11–24 record.

Before the game, Bob blows into the locker room and hands each player a slip of paper. He asks us to vote right now on whether or not we want John as coach. He tells us to write "Yes" or "No." I look over at John, who appears as shocked as anyone. He starts to say something, then slumps into a chair. *He's gone,* I think. *He will quit after this game whether he gets a vote of confidence or not.*

The players who are disenchanted with John eagerly snatch the papers from Bob and furiously scribble the word "No." Other players agonize before they vote. When Bob tallies the votes, he's holding only one slip of paper voting "Yes" to keeping John as coach.

I don't vote. I tear the paper up.

John sits for a moment, then sighs, claps his hands, and says to the room, "Come on, let's go get 'em," and strides onto the court with purpose.

This game against the Nationals, John starts rookies Tommy Hawkins and Rudy LaRusso ahead of the players he knows don't like him. We stumble through the game, playing even worse than usual. I play terribly. I feel as if I'm half there. I try to concentrate on the game, but I keep thinking about John and how intensely and enthusiastically he coaches, how hard he tries. I believe he deserves better. We've lost a lot of close games this season, three or four in overtime. We lose to Syracuse, 109–98, in front of a sparse crowd who braves the frigid temperatures and endures our poor play.

After the game, I shower and change quickly and then wait for everyone to leave. I look for John and find him clearing out his office. After a minute, he comes out, holding a box with his belongings. He manages a thin smile.

"So," he says. "I resigned. I'm leaving before I get fired." He taps his forehead. "Always one step ahead, right, Edge?"

"John." I swallow, and then correct myself. "Coach—"

John holds up his hand. "It's okay."

He waits a long beat and then, as if I'm the one who just received bad news, rests his hand on my arm. "Edge," he says. "I'm sorry."

"It wasn't your fault," I say.

"Well," he says, "I didn't help."

Bob Short replaces John Castellani with former Lakers great Jim Pollard, the "Kangaroo Kid," who, along with George Mikan, led the Lakers to five championships. We'll win a game or two when Jim first takes over as coach, and then we'll go into a disastrous nosedive and lose eleven of our next twelve. I like Jim, and the other players like Jim, but he's no better or worse than John.

It's not about the coach.

It's never about the coach.

• • •

January 17, 1960.

We lose 135–119 in a Sunday afternoon blowout to the St. Louis Hawks. I score 43, but it doesn't matter. We've played four games in four days and lost them all. Since Pollard has taken over as coach, we've dropped five of seven games. We're mired in a slump and nobody can figure out what's wrong.

"We're stinking up every arena," Hot Rod says.

"We're a model of inconsistency," Tommy Hawkins says in the St. Louis airport terminal as we wait to board our flight back to Minneapolis. Unlike other teams, we don't fly commercial—we fly in Bob Short's personal DC-3, a behemoth that looks like a tank with wings. Bob insists that the plane is safer and more reliable than any of the big airlines' planes. That may or may not be true, but I know it's cheaper for him.

This evening, we wait near the food area for our pilot and copilot to arrive. Our plane, now scheduled to take off at 8:30 p.m., has been delayed several hours due to weather. A heavy snowstorm has finally stopped, reduced to a smattering of snowflakes dusting the terminal's windows.

"As long as we have a few minutes," Jim Krebs says, "might as well look into the future."

"Oh, no," I say as Krebs sets up his Ouija board on the table next to us.

"Who wants to help me ask Ouija a question?" he says.

"I will," Larry Foust says. Larry, who likes to medicate himself with alcohol before he flies, has already made several trips to the airport bar. He places his hands on the board, across from Krebs's. The two men close their eyes.

"Ouija, should we get on this plane?" Krebs asks.

"Okay, you know what?" I say, kicking back my chair and getting to my feet. "I'm going for a walk."

I stroll through the terminal, pick up the late edition of the newspaper, stop to get a shoeshine, then sit at the bar and order a glass of white wine. As I wait for the bartender to pour my wine, I watch the snow through the window. It's begun to fall faster and has started to stick to the wing of a plane I see parked below. I finish my wine and head back to the team. I find a seat and open the newspaper. I realize then that the guys have gone quiet.

"What's going on?" I say.

Krebs shifts in his chair. "The Ouija board said we shouldn't fly tonight."

Larry, his eyes closed, slurs as he says, "Ouija said the plane would crash."

I snap the newspaper shut. "You're not serious."

"It also said I'm going to die before the age of thirty," Krebs says.

"Well, hey, you might hit both of those tonight," I say.

"I did not move my fingers," Foust says. "The Ouija board moved my fingers, I swear."

"You've been drinking," I say.

Tommy clears his throat. "I watched the whole thing. Larry didn't do anything."

I look at him. "You don't really believe this, do you?"

Tommy shrugs. "I'm open-minded."

"It's a *game*," I say. "For kids. You buy these things at a toy store."

At that moment, the pilot, Vernon Ullman, and his copilot, Howard Gifford, walk toward us. The instant they show up, I feel better. Captain Ullman flew in both World War II and Korea and handpicked Gifford.

"I think we should take off now," Ullman says. "The snow's starting to pick up again. I want to stay ahead of this storm."

Shortly, all the Lakers—minus Rudy, who missed the trip because of an injury—and a few other people related to the pilot, the copilot, or Bob Short have settled into seats on the DC-3. I sit next to Tommy, in my customary seat near the back of the plane. We take off right at 8:30 p.m., nosing into a steady snowfall. I close my eyes, the roar of the old airplane's generators hammering inside my head. I picture Foust and Krebs, their hands gliding over the Ouija board, and mutter under my breath, "Ridiculous."

As soon as Captain Ullman levels off the plane, I get out of my seat, set up our improvised card table in the front aisle of the plane, and bring out our poker chips and two decks of cards. The five or six of us who play poker settle in for our usual game of seven-card stud. I shuffle one of the decks, finishing with my usual flourish, and we begin to play. When the cards come around for my deal, I snap two cards in front of each player, then deal one up, offering my usual running play-by-play chatter. I scan the up cards, searching for the player with the highest card to begin the betting.

"Well, look at that," I say, tapping the card in front of me. "Ace of hearts. High card. That would be me. That calls for two dollars."

I toss a two-dollar chip into the center of our makeshift table.

The lights flicker and dim.

The card players freeze.

The lights come back up.

The card players relax.

The lights flicker and dim again.

Then the lights go out.

A hush falls over the cabin.

In the pitch dark, I realize that the roar of the engines that previously blocked out all other sound has ebbed, cut out. I hear only the intermittent flapping of the propellers.

In the dark, a voice announces, "Return to your seat."

I stand, lurch, bang into someone seated on the aisle, and mumble,

"Excuse me," pausing for a few moments to allow my eyes to become accustomed to the darkness. As I weave my way back toward my original seat, I pick out Tommy, who stares ahead, his mouth slightly open, his eyes blinking. I can almost smell his fear. I drop into the empty seat across the aisle from him. We sit without speaking, the cabin dark, the propellers whirring.

"We've lost electricity," I say.

"I don't like this," he says.

The plane starts to climb. I lean forward in my seat and grip the armrests. I understand instinctively the pilot's plan. "He's trying to get above the storm," I say. I also know that, since the DC-3's cabin is unpressurized, we can climb only so far before the temperature drops dramatically and we find it difficult to breathe.

The plane continues to ascend.

Then the plane suddenly veers left and falls abruptly. It feels as if the bottom has dropped out. My stomach falls with it.

A few passengers gasp. A few scream.

For the next hour, the plane reels like a drunk, dipping right, listing left, and then once again, the only sound the clacking of the propellers, the plane begins to rise, this time ascending slowly. Without warning, the temperature in the cabin drops viciously. It feels like the inside of an icebox. I grab my coat and pull it up to my chin. As the cold slices into me, I start to pant—I can't find my breath. Then I feel my throat constrict. I gag, force a swallow, and eventually get control of my breathing.

In front of me, a child whimpers. A voice cuts through the darkness, soothing him. I'm struck by the calm in the cabin. I hear no panic. I sense a kind of resolve. I mouth the words:

We're going to make it.

And then, somehow, I let go.

I fall into what some call a Zen state. I blot out time. I eliminate space. I extinguish all emotion. I force myself not to think. I allow myself to put my life in the hands of Captain Ullman. I give him my sacred trust.

We're going to make it.

I believe that.

I know that. I do not — incredibly — experience one second of fear.

I don't know how long we stay flying higher than we should. I will later learn that in the four hours we were airborne, we several times climbed higher than fifteen thousand feet, considered the DC-3 danger point. At one point, Captain Ullman comes out of the cockpit. He stands in front of us, shivering from the cold, his goggles perched on top of his head. He tells us that we have thirty minutes of fuel left. He has maneuvered the plane over some farmland and he is going to land the plane safely on an open spot on the ground.

We're going to make it.

I get out of my seat, sit on the floor in the very back of the plane, and loop my legs and arms around the two seats in front of me. Then I close my eyes.

I hear people rustling around the cabin.

I hear people praying.

I hear some people laughing, a release.

I think I hear, although he later denies it, Dick Garmaker, who's an insurance agent in the offseason, selling people life insurance policies.

Captain Ullman drops the plane to only a few hundred feet above the ground. The propellers flutter and the inside of the cabin rattles as the plane descends. He's looking for a place to land.

Then the plane climbs again, a grid of flickering lights from below illuminating the inside of the cabin. Ullman has brought the plane above the town of Carroll, Iowa, searching for an open space in which to land. The Carroll police, aware of the aircraft in severe distress, have phoned residents and asked them to turn on all of their lights, hoping that the pilot will see the *airport* below. But Ullman doesn't see it. The sudden grid of lights and mass of houses scare him. He has no idea that Carroll, Iowa, *has* an airport. He has no clue that the town below *is* Carroll, Iowa. He doesn't even know we've flown above Iowa. Ullman and Gifford really have no idea where they are. For four hours they have been flying without the use of any instruments. They have been navigating the aircraft by following the stars.

The plane rises and then Ullman banks the plane sharply and again starts to descend. He brings the plane down even lower, this time to about two hundred feet above the ground. He and Gifford frantically scan the land below the belly of the plane, looking, *looking*—

And then they see what they have been searching for: a long patch of earth packed with snow. A cornfield.

Ullman goes for it. He shuts down the speed, lowers the wheels, and eases the aircraft down.

We hit the ground with a rumble and then a thud, and then we rush, roll, and bounce for a hundred feet until we jerk to a sudden and complete stop.

For a count of ten, silence. Nobody speaks. Nobody breathes.

And then all of us, all at once, cheer.

As applause and roaring erupt around me, I let go of the seats in front of me and stand. Red and blue flashing lights flicker through the cabin, coming from every direction. Police cruisers and fire engines have converged, fanning out and surrounding the plane. Dazed, my mouth dry, I wriggle into my overcoat, button up, and follow a few of the guys through the side door. We scurry down a ladder the fire department has leaned against the plane, the frigid night air biting our faces. I come to the bottom rung of the ladder, step off, and plunge into a foot of snow.

I see a wall of police cars and fire engines and what looks like an advancing army of uniforms and reporters with cameras. Past them, at least twenty cars line up along the edge of the field, the cornstalks thin and snowcapped. I take another step or two and see the remains of a barbed wire fence wrapped around the rear wheel of the plane—we must have wiped it out during our landing. I walk a few feet farther and come to an oversize black station wagon. A hearse. A man dressed in a black overcoat stands next to the hearse's open door and smiles. I start to smile back and then I stop. "Are you—?"

"Guilty as charged," he says. "I'm a mortician. The town undertaker."

"First one on the scene," I say.

"I live nearby," he says, in apology, and then he laughs. "I thought I might have some business tonight."

"Sorry to disappoint you," I say.

Behind me, Hot Rod, wearing his usual light jacket, scurries down the ladder and jumps into the snow. "I live to love again!" he shouts.

He rolls in the snow, stands, forms a snowball, and throws it at Slick Leonard, who's lying on the ground, either in prayer or passed out. Slick suddenly leaps up, cradling a snowball of his own, and charges Hot Rod. He slings his snowball at Hot Rod, who slaps it away like Bill Russell, and the two of them start pelting each other with snowballs. Then Hot Rod and Slick go after Garmaker and Foust—four kids in a snowball fight.

Finally, all twenty-two passengers exit the plane, everyone unharmed, most of us too stunned, tired, and relieved to say much. We get into cars and head to a motel where we will spend the rest of the night and regroup, then we'll probably take a train back to Minneapolis. Coach Pollard, the last person off the plane, double-checks that everyone has been accounted for. He realizes then that all of the cars have been filled and that he alone has been left out of the caravan to the motel.

"I'll take you. I got plenty of room."

The mortician.

Pollard looks at him and hesitates.

"Don't worry," the undertaker says. "You can ride up front."

• • •

A week later, back in Minneapolis, we lose to the Knicks 115–104, our fifth-consecutive defeat. I score 30 and Dick Garmaker pops in 29, playing his best game of the year, his third or fourth excellent game in a row. The next day, Bob Short trades him.

He sends Garmaker to the Knicks for backup center Ray Felix and $25,000 in cash. I hate to see Garmaker go, but I get why Short made the deal. We need Felix for his size and relative quickness when we go up against big, agile centers, in particular Wilt. And we all know that the Lakers franchise is leaking money. I don't know how severe a cash flow problem Bob faces, but someone tells me that Bob needs the $25,000 to meet this month's payroll. To me, that's severe. I'm not losing sleep about the team's financial state the way I imagine Bob does, but a blind man could see the signs. We're playing poorly, and nobody's coming to the games. The team doesn't talk much about it, but we all share the same concern. Finally Pollard and Hot Rod say something to Short. He tells them he's working on an escape route.

Short's escape route runs from the Midwest to the West Coast. Bob took special note of the crowds when we played games last season in San Francisco and Seattle. Maybe it's the milder weather or maybe it's because basketball is the new, new thing on the West Coast, but we draw big crowds who not only come out to see us, but love what they see. Again, we're 13–31, not exactly a great basketball team, but we don't draw crowds that large or enthusiastic in Minnesota when we're *winning*.

We all love San Francisco, but Bob has his sights on Los Angeles. He schedules us to play on February 1 in the L.A. Sports Arena. We will take on the Philadelphia Warriors, a game some bill as a battle between the likely 1960 Rookie of the Year, Wilt Chamberlain, and the 1959 Rookie of the Year — me.

We arrive in Los Angeles several hours before tip-off. We check into our hotel and then Hot Rod and I decide to check out the beach. As we squeeze into the back seat of a cab, the driver stares at us. "Are you guys basketball players?"

"No," I say. "Jockeys."

"Step on it — we're late for the fifth race at Hollywood Park," Hot Rod says.

The cabbie laughs and then decides to serve as our personal tour guide. He drives us through the city, from downtown to the beach, pointing out the sights, taking a side trip up to Hollywood Boulevard, cruising over the Sunset Strip, then sauntering down Rodeo Drive to show us the mansions and high-end shops in Beverly Hills and finally heading west and arriving at the end of Wilshire Boulevard, in Santa Monica. He waits for us as Hot Rod and I walk slowly along the bluff overlooking the Pacific Ocean. We walk past dozens of palm trees, their fronds rippling and swaying in the gentle breeze. The noontime heat feels warm and soothing, unlike the oppressive humidity I grew up with all those summers in D.C. Then I think about the evil winters we endure in Minnesota.

"It's February *first*, baby," Hot Rod says. "It was ten degrees when we left."

Rod and I pull up to a short fence at the edge of the bluff, overlooking the Pacific Coast Highway. Beyond that lies a wide, smooth Santa Monica beach that kisses the Pacific Ocean, whose pale blue expanse disappears into the horizon. If you can fall in love with a place, then I have fallen in love.

"Man," I say. "Can you imagine if Bob moved the team here?"

"It's like this twelve months a year," Rod says.

"I could get used to this," I say. "If I tried really hard."

Hot Rod laughs. "Amen, baby. *Amen.*"

We lose the game to Philadelphia, 103–96. I score 36, but Ray Felix can't contain Wilt. He puts in only 23, seemingly having more fun passing the ball. But the number that Bob focuses on is 10,100: the number of people who attend the game. The next day the local newspapers, including the African American paper the *Los Angeles Sentinel,* drool over both Wilt and me, practically begging Los Angeles to land an NBA franchise. I find Hot Rod sitting in the hotel lobby reading the newspapers. We're about to catch a shuttle to the airport and fly — *commercial* — to play the Celtics in Boston. Hot Rod looks up when he sees me and slaps the paper. "The *L.A.* Lakers," he says. "What do you think?"

"I like the sound of that."

"Amen, baby," Rod says.

• • •

We come back to Los Angeles three weeks later to play back-to-back games with the St. Louis Hawks. We win the first game and drop the second, in front of big crowds. Soon after we return to Minneapolis, rumors swirl. There seems little doubt that Bob Short wants to move the team to start the 1960–61 season. But a decision like that always involves money and politics. I hear that the NBA balks at first, suggesting that Short consider a move to an East Coast city—Baltimore or Pittsburgh—neither one my first choice now that my love affair with L.A. has bloomed. Then I hear that Bob may be debating between San Francisco and Seattle, cities I like, but both of which have geography I don't love—steep hills that make them treacherous to navigate—and iffy weather. I realize I crave warm weather. I want 24/7 summertime. I want L.A.

Then I hear that Los Angeles power brokers—the people Bob needs on his side to make the transition from Minneapolis—and the NBA brass are split. Some see the Lakers with me as their centerpiece the best fit for L.A. Others want Wilt and the Warriors. The opposing factions feud, putting forth passionate arguments and getting into public relations battles, to the point that someone makes a mock-up of me in a fake Los Angeles Lakers uniform and prints it in a newspaper. I like it. I want it.

But I can't control what's going to happen, so I decide to let it go. Plus, I don't want to think about moving to L.A., because I'm afraid I might jinx it. I'm sure Krebs has always asked his trusty Ouija board about the team's fate, but I don't want to know what it answered, not after that plane flight. I do know that the decision has to be put on a fast track, because we're already into the first week of March. We will only have a few months to pull up stakes in the Twin Cities and settle in L.A.

With six games left in the season, I concentrate on basketball and Ruby. Ruby, looking and feeling as if she could give birth at any moment, stops attending home games. Maybe because I've become one jangled nerve ending and my constant nervousness juices me up, we win four of those final six—both losses coming by one point, one of them in overtime—and I average almost 37 per game.

A considerate child from day one, my son, Alan, arrives on March 8, an off day. That day I pace in the hospital waiting room until a nurse bursts in and announces, "Congratulations. You have a son." In those days, doctors kept dads out of the delivery room—not a bad idea in my case, since I probably would've gotten in the way or passed out.

The day after Alan's birth, although I'm exhausted, I score 31 and we lose. The day after that, March 10, I manage to get some sleep and we win our final game of the season, 133–108 against the Knicks. I score 37, and Dick Garmaker, now a guard for New York, puts in 18 for the Knicks. After the game, Dick reunites with his partying pals Slick and Hot Rod for a night of I-really-don't-want-to-know. They do plan well, allowing themselves a full day to recover. We have an open date before we take on the Detroit Pistons in the first round of the playoffs, despite our finishing third in the division, with a 25–50 record. With six of the eight teams in the league getting in, the postseason truly is a new season.

· · ·

Nobody expects us to do much in the playoffs, but I do.

We begin the short, best-of-three series in Detroit on Saturday night, March 12, and at first we really don't do much. We flail around for three quarters, and then I catch fire in the fourth and we obliterate the Pistons' nine-point lead. We stun them, 113–112, their loud crowd becoming abruptly silent: six thousand statues. I pour in 40, and several other guys score in double figures. The Pistons never recover. We play the second game of the series in Minneapolis and carve them up by 15. I score 25, but Frank Selvy gets hot from the outside and leads us with 30, and just like that, to the surprise of the pundits and so-called experts, we advance to the semifinals to take on the always testy St. Louis Hawks in a best-of-seven series that turns into a war.

We play the first two games in St. Louis and I'm way off in game one, scoring a frustrating 19, scrounging for every point, missing a ton of easy shots. We get our butts kicked, 112–99.

Game two is a different story. I'm determined to steal the second game. I go for 40, and we shock the Hawks, 120–113. We head back home to Minneapolis tied at one game apiece.

Game three quickly becomes a low-scoring, back-and-forth bloodbath. We look overmatched in the first quarter, playing tense. I throw the ball away a couple of times and I feel as if we're completely out of sync. The quarter ends with us trailing by 11.

We loosen up and fight back in the second quarter. I score on a couple of quick jumpers and a floating-in-the-air layup, Rudy hangs tough against Cliff Hagan, and at the half we trail by only one. The battle rages in the second half. We pull ahead by two, fall back by six, fight back, tie them up, go back down by four, and then, midway through the fourth quarter, I get hit with my fifth foul. I back off Pettit the rest of the way, desperately trying not to foul out. I manage that, but he keeps coming at me. He scores 35 to my 27, Hagan adds 30, and the Hawks nip us 93–89.

That game just pisses me off.

I don't play well angry, and I scuffle through game four. We go down by 12 as the third quarter ends. I sink onto the bench, staring straight ahead. I put up a wall around me. Nobody talks to me. Nobody comes near me. I simmer. We've had a rocky season: We've won a total of twenty-five games. We've gone through two coaches — John, who deserved a better outcome, and Jim, a great guy, but the definition of "interim." We've seen the crowds dwindle and lose interest. We all nearly died in a plane crash. We've sweated out our owner's ability to pay us every month. And we keep hearing that we're moving somewhere: the West Coast, probably; Los Angeles, hopefully. I want to salvage this season; we can do that. We just have to beat the Hawks. Something about them fires me up. Maybe because Cliff Hagan and Bob Pettit, two of the best forwards in the league, seem to challenge me more than anyone other than Tommy Heinsohn. I don't know. I just really want to beat them.

I play with total abandon in the fourth quarter. We slowly catch up to St. Louis and finally pull ahead to stay, late in the period. We win 103–101. I put in 37 and practically dance into the dressing room. The series tied, we head back to St. Louis.

Game five starts out so badly, it almost feels as if we haven't gotten off the bus until the start of the second quarter. In the first quarter, we score exactly 14 points and go down by 12. The way we've played, we're lucky to be down by *only* 12. Then—click—we flip an imaginary switch and turn it around in the second quarter, outscoring the Hawks by 12 and knotting the score 47–47 at halftime. In the second half, we go toe to toe, battering each other, trading baskets and bone-crushing body shots. We end regulation tied at 103. In overtime, I score two quick baskets and then Hagan gets called for his sixth foul. Thirty seconds later, Pettit commits his sixth foul. With those two on the bench, we step on the gas and blow the Hawks out, 117–110. I finish with 40 and we take a 3–2 series lead.

Now I think, *Whoever wins this game wins the series.*

I feel that so strongly.

Except—I'm wrong.

Two days later, we go home to Minneapolis for game six and we fall apart. We can't handle their front line. They pound us inside. Pettit scores 30, Hagan and their bulky center Clyde Lovellette each pump in 29, and, despite my 38, St. Louis swamps us 117–96. The Hawks run over us in the first quarter and never let up. With the series tied 3–3, we return to St. Louis for the final game, winner take all.

We lose 97–86. We never really make it much of a game. I manage 33 points, but the Hawks smother everybody else. In the locker room after the game, we change in silence. Even Hot Rod goes shockingly mum, except to mutter under his breath, "We had them," before slamming his locker door. I feel stunned. I can't figure out why we collapsed and lost the last two games so badly. We gave this series away, I know that. Maybe we just ran out of gas. Maybe we just didn't have enough to finish them off. Or—

Hell, we lost. That's what happened. We—*lost.* We shouldn't have, but we did. I finish getting dressed, stuff my gear into my bag, and quietly close the locker door. I scan the room, taking a long look at my teammates. I have such affection for these guys. We've formed a family. We've gone to battle every night for five months. We have each other's backs. We'll keep fighting—together—but everyone in this room knows that we face the inevitable.

Things are about to change.

We return to Minneapolis, and for a day or so I walk through my life feeling on edge. The league announces the All-NBA teams and individual honors. Statistically, I've done even better than in my rookie season, averaging 29.6 points and 16.4 rebounds a game. I join Bob Pettit, Gene Shue, Bob Cousy, and Wilt Chamberlain on the All-NBA First Team. But the season leaves a sour taste.

One night, a week or so after the playoffs, with Alan asleep in his crib, Ruby and I sit on the couch in our living room, the blue light of the TV bathing us. An image on the screen pulls me in and I focus on the two figures before me: a sportscaster and an athlete in midinterview after a game. The sportscaster shoves a microphone into the athlete's face and asks, "How did you feel when—?"

I suddenly imagine myself standing in that athlete's place and the sportscaster asking me, "Elgin, how did you feel when you scored 64 points to break the all-time NBA record?"

Chris, I felt fine. I felt good. But I only wanted to win the game. My teammates kept feeding me because I was open. We play as a team. That's what it's all about: the team. Just nice to get the win.

That's what you say. Sports-speak. The art of the cliché. But I'd be lying.

For a stretch of time—a quarter, maybe even an entire half—I felt superhuman. I felt magical. I felt that I could do anything out there.

But I honestly didn't think about it. I didn't.

You don't think—you *do*.

Athletes don't plan. We don't strategize. We do the opposite: we shut off our minds. We act on instinct.

Somebody asks me once—a reporter—what it feels like to be great. I have no idea. I never think about it. I just . . . play.

"But how does it feel being so much better than every—"

I cut off the question. "I don't know. I never think about anything like that."

If I'm so great, why did we win only twenty-five games in the regular season and lose to the Hawks in the playoffs after going up three games to two? I can't explain it. *I don't think about it.* I don't want to think about it.

"I want to move to L.A.," I say to Ruby, the words spilling out.

Ruby turns to me. "Why? I like it here. I thought you liked it here."

"I do. It's just . . ."

I stop. I blink at the TV. The sportscaster and the athlete have dissolved into two cowboys sitting on horses, proclaiming their love for Marlboros.

"What?" Ruby says.

"Maybe we all need a fresh start. Us, the team, everybody. Change of scenery. Something with palm trees."

7

HOLLYWOOD (ACT I)

THE HEADLINE NEARLY BLINDS ME.

LAKERS TRADE BAYLOR TO HAWKS FOR $200,000 CASH AND A PLAYER TO BE NAMED LATER.

I'm shocked.

The Hawks would pay $200,000 just for me? That's more than Bob Short paid for the whole Lakers franchise.

Then my mind goes straight to basketball. I picture my new front-court—Bob Pettit, Cliff Hagan, and me. The three of us would tear the league *up*.

Almost as soon as I read the headline, my phone rings. Bob Short.

"Is this a serious offer?" I say.

"Completely serious," Bob Short says. "They know I'm strapped for cash."

"It's crazy," I say.

"Unheard of," Bob Short says. "I turned them down."

The silence between us goes on for so long I think the phone has died.

Finally, I say, "So, you said no—"

"Elgin, they're acting like I'm having a going-out-of-business sale. I'm not going out of business. Far from it. We're just relocating."

"To Los Angeles?" I say hopefully.

"You got it. The hell I'm trading you. We're moving *because* of you. You're our centerpiece. You're the franchise."

Of course, I don't want to be traded. I especially don't want to be traded to St. Louis. Bill Russell and several other black players around the league constantly remind me how some of the St. Louis fans taunt them during games, slinging racial slurs at them mercilessly. But I can't help kidding Bob Short, even during what I know to be an anxious time. "I don't know, Bob. Pettit, Hagan, me — we would do some serious damage."

"There's only one ball," Bob Short says.

"It's a lot of money. You sure you can afford to pass it up?"

"You'll need a new wardrobe in L.A." Bob sniffs. "Go heavy on polo shirts and Bermuda shorts."

"With your legs," I say, "you may want to hold off on the shorts."

He sniffs again. "I never said *I* was relocating."

· · ·

As the players prepare to move to Los Angeles, Bob Short begins re-tooling the team, confident he can run the franchise long-distance. He entrusts our general manager, Lou Mohs, to handle all the day-to-day operations. While Ruby and I are packing up and having a yard sale, Bob Short takes step one with the team. He sits down with our coach, Jim Pollard.

Jim has asked Bob Short for a meeting to discuss his future. Much to Bob's surprise, Jim asks for a raise and a long-term contract. In order to function at his best, Jim says, he needs job security. Bob listens to Jim's compelling argument and then fires him.

Like a chess player, Bob has been thinking two moves ahead. He and Lou have long coveted University of West Virginia coach Fred Schaus. Fred, a former NBA forward, has become one of the most successful college coaches in the country, leading West Virginia to six consecutive NCAA tournament appearances and making it to the final in 1959, where they lost the championship by one point. In addition, he coached both Hot Rod Hundley and Jerry West, a two-time All-American forward, who finished this past college year averaging 29 points, 16 rebounds, and four assists a game while shooting 50 percent from the field. I follow college basketball and I've seen West play. He can definitely score, but even more than that, he can rebound and defend, and he's ultracompetitive. He looks like a choirboy but plays like an assassin.

The Cincinnati Royals have the first pick in the upcoming draft and will no doubt go for the hometown favorite, University of Cincinnati all-everything guard Oscar Robertson. That will open the door for the Lakers to grab Jerry West with the second pick. When Mohs and Short dangle in front of Fred Schaus the carrot of once again coaching both Hot Rod and Jerry West, Fred jumps at the chance. He signs a multiyear contract.

I tell Hot Rod the news. His face flushes, the way it does when he feels angry, or upset, or has a good poker hand. "*Schaus?* He's our new coach?"

"You don't seem thrilled."

"Well," Hot Rod says, dragging out the word to about four syllables. "Fred's an acquired taste. He's a good coach. He's just extremely . . . I'm looking for the right word . . . *no-nonsense*."

• • •

We move to Los Angeles over a frenetic two weeks. Ruby and I find a duplex in midtown, an easy drive to the new, modern Los Angeles Sports Arena, where I'll be playing basketball, and a short distance from the University of Southern California, where Ruby will be going for her degree in education. In early June, once everyone has gotten settled, Fred Schaus gathers the team for an informal practice to get to know us —and for us to get to know him.

I stand with Tommy Hawkins, across from Hot Rod and Jerry, who keep their eyes fixed on the floor as Fred goes through his personal rules and regulations and explains his coaching style. I get what Hot Rod meant by "no-nonsense." Fred comes on strong. He sounds more like a drill sergeant than a basketball coach. After a while, I tune him out. Hot Rod, I notice, keeps his head down and his eyes averted, but I see that his shoulders have started shaking and that he's trying not to laugh. Jerry, though, seems lost in his own world. He shifts his weight and keeps his hands clasped behind him. He looks unhappy and uncomfortable. When Fred finishes speaking, he puts us through a few drills. Jerry, I see, has skills—quick feet, a nose for the ball, and he can shoot the lights out. I also see that he says almost nothing and smiles not at all.

Later, Lou Mohs tells us that part of our responsibility over the summer, in addition to staying in shape and attending practice, will be community outreach. We're the new kids in town. Nobody knows us, and we're not sure how well the L.A. sports community will embrace us, or whether they will embrace us at all. Fred wants people to get to know us. He assigns each of us to a different basketball camp or clinic somewhere in the vast grid of towns and neighborhoods that make up greater Los Angeles. I find out that I have been assigned a children's basketball camp that goes on for an entire week. Then I see that I'll be working the camp with UCLA coach John Wooden, the future legend, who by this point has not yet begun his record run of ten NCAA championships. I don't actually spend much time with Coach Wooden, but when I do meet him, briefly, he lights up.

"I wish I had you in college," he says. "You're a team player."

From John Wooden, high praise.

• • •

I gravitate toward Jerry West. Coming from a small town in West Virginia, he seems overwhelmed by the size and pace of L.A. He and Hot Rod both went to the University of West Virginia and played for Fred Schaus, but that may be exactly everything they have in common. Hot Rod, Slick, and a few of the other guys seem to be on a quest to cover every inch of L.A., one bar at a time. Jerry doesn't fit into their fast, bar-hopping lifestyle. Hot Rod lives to have fun. Jerry lives for basketball.

One day after practice, when everyone else has gone, Jerry and I start shooting around. Before we know it, we're playing one-on-one, just for fun. Well, I'm playing just for fun. Not sure Jerry can do that. First of all, when we face each other head to head, Jerry, though listed at six foot two, is closer to six-four. Second, Jerry goes after you. As I begin my dribble, he gets right up on me. I lower my shoulder, bump him a little bit, make my first move, and ward him off with my arm. He holds his ground, refusing to surrender one inch. I give him a fake. He goes for it. I streak by him and beat him on a drive to the basket. He slams his hands together in an irritated clap, then dares me to try that same drive again. I do, and this time he stares at my footwork. I fake a move to the hoop, step back, and bank a jump shot off the backboard. He rests his hands on his hips and looks up at the backboard, eyeing the spot where the ball kissed off the glass.

He's challenging me, I think.

Then I realize: he's not. He's studying me. He's trying to learn from me.

A few days later, at the beginning of practice, Fred Schaus summons Jerry to him. They stroll to a corner of the gym, Fred talking and gesturing the whole time, Jerry listening intently. After the conversation, Jerry comes up to me, a glum look on his face.

"What was that about?" I ask.

"He doesn't want me to play forward. He wants me to play point guard."

I have no doubt that Jerry can learn to play guard, but with Hot Rod, Slick, and Frank Selvy, we seem overloaded at that position. But Schaus gets frustrated and even annoyed at Hot Rod's on-court antics and unpredictability, Slick fares better coming off the bench, and Selvy, an excellent shooter, doesn't handle the ball that well. Jerry, even though he's tenacious as hell, wouldn't be able to dislodge Rudy, Tommy, or me from our forward positions. We all have too much size.

"He may be right," I say. "You have great vision. Point guard might even be your natural position."

Jerry's expression goes dark. He seems to speak from far away. "He says he won't start me until I'm *ready*."

"You'll be ready by our first game," I say.

"He doesn't think so." He turns away and picks up a basketball. He dribbles a couple of times and then banks in a jump shot off the backboard. Guy's a quick study.

"Did he say when you would start?"

"No," Jerry says, his voice low.

I'm sure that Jerry has started every basketball game he's ever played. Not starting, even though he's learning a new position, must feel like a demotion. It must feel like a loss.

As I will learn, as he himself will confess, Jerry West can't stand to lose.

• • •

I need a car. I want something reliable, comfortable, and not too flashy. I buy a pink Oldsmobile 98 convertible.

I don't realize the car is pink . . . I'm color-blind. I think it's beige, or light gold. The salesman never corrects me or double-checks that I want to buy a pink car. He's probably so shocked that somebody actually wants the only pink car in the showroom that he doesn't say anything that might blow the deal. I don't find out that I've bought a pink car until I'm on my way home.

With the top down and the radio turned up, I drive from Beverly Hills across Santa Monica Boulevard through West Hollywood, a mostly gay section of Los Angeles.

I stop at a red light, drape my arm across the top of the bench seat and bob my head to a Sam Cooke tune. The passenger door suddenly flies open and a guy wearing a tank top and short-shorts catapults himself into the car and slides right next to me. "I'm going home with *you*," he says.

I'm too stunned to speak. The light turns green and the guy flies out of the car as fast as he came in. "Hot pink," he shouts, rubbing the side of the car and pursing his lips. "That's my color."

Ruby loves the car when I show it to her, but she looks confused.

"I know," I say. "It's pink. I found that out about twenty minutes ago."

I keep the car, and a few weeks later I buy a second one.

Ruby comes with me to pick out the color.

. . .

At the end of September, the night before we play the first of several exhibition games for charity, Ruby and I watch the first-ever televised presidential debate, between Democratic candidate John F. Kennedy, an upstart senator, and Republican nominee Richard M. Nixon, the current vice president.

Our country has begun to face more turmoil associated with civil rights. In March, four young black people staged a sit-in at a segregated Woolworth's lunch counter in North Carolina, insisting that they should be served and that segregation should be abolished. Their sit-in reminds me of my refusing to play in Hot Rod's hometown in West Virginia. Since my rookie year, three years ago, I have seen more and more black players come into the league. *The times they are a-changin',* the song says, and for the better. I want to vote for the presidential candidate who supports civil rights.

In the debate, Kennedy appears relaxed and poised. Nixon seems more nervous than the car salesman who sold me the pink Oldsmobile. Whenever the camera goes close on him, he blinks rapidly, making him appear shifty. As the debate concludes, I find myself fascinated by the amount of perspiration that has accumulated above Nixon's upper lip. It looks like a lake.

"I've never seen anyone sweat so much," I say to Ruby. "He's sweating more than I do in the fourth quarter."

In October, we play an exhibition game against the Celtics in Anaheim, home of Disneyland. After the game, as usual, I meet Bill Russell for dinner. This time I bring along Jerry West. The two future rivals and Hall of Famers immediately click. The next day, I spend some more time with Bill and he tells me how much he enjoyed meeting Jerry.

"He's a kindred spirit," Bill says. "I saw that on the court, the way he plays. He reminds me of me."

"Because he's such a good defender."

"I meant because he's so intense."

. . .

We start the season playing two games on the road, losing first to Cincinnati, 140–123, and then to St. Louis, 112–96. We come back to L.A.
to play our first game in our new home, against the New York Knicks.
Despite all our efforts over the summer to introduce ourselves to the Los
Angeles sports community and all the supposed excitement associated
with our first home game, we play to a half-filled arena. And we lose
111–101, our third loss in a row. The next night, we play the Knicks
again and win our first game of the year, 120–118. I score 36 and Jerry
adds 18 off the bench.

After that, we play better, winning every other game and bringing
our record to 4–6. Fred increases Jerry's minutes every game and often
keeps him on the court for most of the fourth quarter. Still, he doesn't
start him. Worse, he rides Jerry hard in practice and during games, getting on him when he makes a mistake. Jerry chooses to keep quiet and
seethe. I tell him that I think Fred is just toughening him up.

"He will start you soon," I say. "He has to. You can learn on the
job."

Jerry says nothing. But I can see his eyes riveted on Hot Rod, who
sends a behind-the-back pass sailing past Rudy LaRusso and into the
first row of the Sports Arena. Jerry frowns.

"Once Fred puts me in the starting lineup, nobody will ever get me
out," Jerry says. He's not speaking to me. He's making a promise to
himself.

As we hit the road for several games, I spend Election Night, November 8, watching the results of the presidential election in a bar
with some guys on our team. The team seems divided, as does much
of the country, but John Kennedy wins in an upset. I can't say I feel
happy, but for the first time in my life, certainly my adult life, I feel
hopeful.

• • •

We play our eleventh game of the year in Providence, Rhode Island, against the Celtics. Even this early in the season, everybody on the team feels on edge. Fred, a stickler for discipline, doesn't love that Hot Rod and his posse hit the bars after every game. Fred toys with imposing a curfew. He floats the idea to me, but I insist that we're grown men and professionals, not college kids, and I know the players will resent a curfew. I tell him that certain guys, and I don't mention names, will treat a curfew not as a rule but as a suggestion. Fred abandons the idea.

We're also on edge because a lot of us feel uprooted. Since we're based in Los Angeles—the only team in the league located west of St. Louis—we find ourselves forced to travel more than any other team. I look at the schedule and see that during a two-week period that includes Christmas and New Year's Day, we play ten games in a row on the road. When we're home, we have to avoid falling into a rut. Again, because we're the only team in the West, teams come to Los Angeles and stay for days. We play the same team twice in a row, even three times, and in one case *four* times.

Fred does impose one new rule. He decides that we should rotate roommates, switching to a new guy every road trip. I understand what Fred's going for—team unity and getting to know someone you may know only casually—but I need my rest. I'm a terrible sleeper and a ridiculously early riser. I do best on eight hours a night, but I'm happy if I get four or five. I tell Fred that I *really* need my sleep. I don't think he understands. I don't want to be disruptive, so I go along with Fred's rotating roommates program without complaining, but I have my doubts.

First Fred pairs me with Rudy LaRusso. I love Rudy. He's intelligent, engaging, fun, doesn't really party or stay out late, and he's very considerate. He also doesn't seem to shower or use deodorant. Maybe I have a supersensitive nose, but his smell permeates the room. Seconds after he gets into bed, he's out cold and I can't sleep.

I start calling him "Musty" because of his odor. I don't think he realizes why I call him Musty, but the name sticks. Then everybody starts calling him Musty without knowing why. They'll find out once they room with him.

After Rudy, I room with Jerry. Jerry is easy, an ideal roommate. He has a curious and analytical mind when it comes to basketball, and we talk basketball constantly. We break down players, we talk plays, we talk shooting, we talk strategy. Except after a loss, when we don't talk.

On those nights, Jerry disappears into his silent — and even he might say — tortured world. He broods. I don't try to breach his silence, and I don't want to say anything to add to his pain, so I keep my distance, allow him to deal with losing the way he has to.

After Jerry, the parade of roommates continues. I'm sure guys complain about me, too. I'm sure I have my own annoying quirks. I can't think of any, but I'm sure I have *some*.

I honestly care only about getting my measly four or five fitful hours of sleep. Having a roommate starts to wear on me. I endure:

A guy who can't sleep without the lights on.

A guy who can't sleep without the TV or radio on, soothing white noise to him, irritating noise to me.

A guy who comes in at dawn, drunk, and then talks nonstop about his night, slurring his words, laughing at the top of his lungs.

A guy who comes in at dawn, noisily passes out on his bed, waking me up, or heads straight into the bathroom and vomits, waking me up.

A guy who asks me to room with somebody else for the night because he has a woman with him.

A guy who snores. Really snores. Like a freight train.

After my seventh or eighth roommate disaster, I seek out Fred Schaus in his office at the Sports Arena. I close the door behind me. "Fred, I love my teammates. I love them as players and I love them as people. But I can't room with anybody. I'll pay the difference between a double room and a single if that's the problem, but I need my own room on the road. I need my sleep."

"Is it that bad?"

"Fred, I'm so on edge right now because of lack of sleep that I may kill the next guy to make sure he's quiet."

From then on, I have my own room on the road.

I'm sure Fred understands my quirkiness, because he's quirky, too.

Against the Celtics in Providence, we streak to an eight-point lead at the end of the first quarter. In the second quarter, we fall apart. The Celtics outscore us by 14 points and we go into halftime down by six, a ridiculous 22-point turnaround. In the first half, Fred keeps leaping off the bench and then slamming himself back down. Every time we screw up, which is often, he stomps both feet.

From then on, I call him either "Thumper" or "the Stomper."

Not to his face.

In the locker room at halftime, Fred loses his mind. He shouts, curses, tries to outline a play on the chalkboard, but the chalk shatters in his hand. He tosses the remaining pieces in the air like confetti. His face purple, he turns to Hot Rod. "You got a cigarette? I left mine at the hotel."

"Sure, Coach," Hot Rod says in his raspy voice, the result of a lifetime of smoking and drinking whiskey. Rod opens his locker, finds his pack of cigarettes and his lighter, taps out a cigarette for Fred, and lights it for him with a flourish.

Fred inhales, tilts his head, blows a stream of smoke to the ceiling, turns to us, and starts screaming again.

In 1960, everybody smokes. Baseball players smoke. Football players smoke. NBA players smoke. Guys on our team smoke. More than half the fans in the stands at NBA games smoke.

I don't smoke. I did for a short time, but I gave it up. Two reasons:

I found that smoking cut down my endurance.

And I can't stand how cigarettes make my clothes smell.

In that game against the Celtics, we fight hard in the second half but still come up short, 131–124. I put in 45, and Jerry, again off the bench, scores 27. Eight different Celtics score in double figures. Our record drops to 4–7.

We bus over to New York to take on the Knicks a couple of nights later.

It would be a night to remember.

• • •

November 15, 1960.

Something's in the air.

You can feel it, a kind of electricity crackling through Madison Square Garden, rippling through the sold-out stands, pulsing beneath every inch of hardwood floor.

As the two teams line up for the center jump to start the game, I shake hands with Willie Naulls, one of the Knicks' forwards and a new friend who's from L.A.

"Take it easy on us, will you, Elgin?"

"I don't know, Willie," I say. "I feel something different tonight."

"Different how?"

"I just feel good."

"Uh-oh. I don't like *that*."

We lose the center jump, Willie knocks in a short jump shot, and we go down 2–0. I take a pass from Hot Rod, dribble twice, spin, soar past the basket, and flip in a reverse layup. After that, I shut everything out, setting up guys when I can, rebounding ferociously, and scoring when I have an opening. This night, though, I can't miss. Every move I make gives me space, and every shot I take goes in. I usually feel pretty good on a basketball court, confident that when I have the ball in my hands, I can do whatever I want—especially score. Not that I always do score when I want to, but I *feel* that I can. Tonight feels surreal. I feel almost as if I'm back at the playground around the corner from Heckman Street in D.C., shooting hoops by myself as the sun sets.

At one point in the fourth quarter, Fred calls timeout. I share a look with Jerry and wonder, *We're up, we're winning—what does Thumper want now?*

"Keep feeding Elgin," Fred says, swatting his clipboard against his leg.

I look at him.

"He has 59 points," Fred says, avoiding my eyes. "Six more and he breaks his own record."

"I had no idea," I say.

"Let's do this for him," Hot Rod says.

The guys feed me for the rest of the game. On each shot I make, the Madison Square Garden full house of Knicks fans applauds and cheers. I know I've scored more than 64, but I honestly have no idea how many more.

We win 123–108. As the horn sounds ending the game, the crowd roars, most of them on their feet, cheering. Willie Naulls runs over to me and starts pumping my hand. "Congratulations," he says, beaming. "You more than doubled me up."

"What?"

"I had 35," Willie says, still smiling.

I still don't process what he's saying. "I had—"

"Seventy-one points," Willie says, my teammates circling around me.

After the game, I leave the locker room with Hot Rod. Feeling a little giddy, I agree to have a nightcap with him in the hotel bar. He hails a cab and the two of us scramble into the back seat.

"You know who this is?" Hot Rod shouts to the cabdriver. "Elgin Baylor, the greatest basketball player in the world. We just beat the Knicks. You got seventy-eight points in this cab, baby. *Seventy-eight points.*"

"That's right," I say. "Hot Rod here had seven."

• • •

We go on a tear on the road after that, winning five of our next six games and lifting our record to 10–8. I stay torrid, scoring literally at will. After pouring in 71 against the Knicks, I score 52 against the Pistons a few nights later at home, and 51 the night after that, in a game we play in San Francisco. In those six wins, I average 48 points a game.

Then we drop two straight and Fred decides to make a change.

"I'm starting," Jerry says, stone-faced, as we line up for our layup drill, finally back at the Sports Arena after what seems like a month. "He just told me."

"He told me, too," I say, eyeing the Philadelphia Warriors at the opposite end of the court. "You nervous?"

"I wasn't until you just asked me," Jerry says.

"It's no big thing," I say. "Oh, did he mention you're guarding Wilt?"

At least Jerry laughs at that.

We all play poorly against Philly and lose three out of four games against the Warriors as Jerry adjusts to starting, trying to find his rhythm. We head back across the country to Boston for one game, turn around and come back home for two more against the Celtics, then hit the road again for the last two weeks of the year. Either a drunk made this schedule or Red Auerbach.

Going into the new year, we play sub-.500 basketball even though Jerry finds his comfort zone and plays every game now the way everybody has predicted: fluidly, smartly, and with exceptional toughness. Most games, the two of us lead the team in scoring, usually putting in at least 50 points between us and often more than 60. The league votes three Lakers to the 1961 NBA All-Star Game: Jerry, a rookie, who has started fewer than half of our games; Hot Rod, who now comes off the bench to spell Jerry; and me.

A few games after the All-Star Game, we catch fire and, in the last long stretch of the season, go 17–12 to bring our regular season record to 36–43 — not great at all, but good for second place in the West.

Just before the end of the season, in early March, riding a three-game winning streak, we go into St. Louis to take on the Hawks, the top team in the West and our likely opponent in the Western Division Finals if we get past the Pistons in the first round. In front of the usual raucous St. Louis crowd, both teams get testy and the game deteriorates into a slugfest. The officials don't help as we get hit with the worst case of home cooking I've seen since that NCAA tournament final matchup against Kentucky. I usually have nothing to say about the officials, but this night I feel like the two referees have put money on the Hawks, because that's how they're calling the game. We play two overtimes and eventually lose an exhausting marathon, 137–136. I score 39, but I foul out, as do Jerry and Rudy. In fairness, the officials foul out a couple of Hawks, too, Hagan being one, but somehow in this foul festival Bob Pettit gets called for only two fouls on his way to 48 points.

A referee calls me for my fifth foul—a phantom foul, occurring only in his imagination—indicating that I slapped Pettit on the arm instead of blocking his shot. The Stomper flies off the bench as if it were an ejector seat, slams his butt back down, and stomps so hard I'm afraid he'll put his feet through the floor.

Meanwhile, my patience, withering all night, dissolves. I walk in a circle to calm myself. It doesn't help. I feel the rage burning. I hear the crowd booing. I hear some racist taunting, and that clinches it.

I speed-walk to that referee. I see his eyes cloud as I approach.

Normally, if I have a question or a complaint and I want to talk to an official, he'll say, "Elgin, just stand here next to me, look straight ahead, and talk out of the side of your mouth. I'll listen to you, but it looks bad if it seems like you're questioning me directly." So that's what I do.

Of course, this will seem ridiculous in later years, when players confront officials on every call, gesturing, complaining, arguing, demanding a detailed explanation. In 1961, players—and fans—have a different relationship with referees.

There will be one occasion when fans boo a particular referee because of a series of questionable calls against the home team. After the game, a fan hops out of the stands and charges this official. The fan isn't aware that before the referee took up officiating, he boxed professionally. The fan swings at the referee. The referee slips the punch and knocks the fan unconscious with a right cross to the jaw.

This night in St. Louis, the night of that fifth phantom call, I don't raise my fist to the official, but I do raise my voice.

"That was not a foul and you know it," I scream. "I got all ball!"

"Elgin, back off."

"You're missing calls all night. You've been terrible!"

"I don't want to tee you up—"

"You're the worst. If I wanted to see a clown, I would've gone to the circus."

And that's how I receive the first and only technical foul of my career.

. . .

March 14, 1961. Our first postseason in Los Angeles.

Since our official introduction to the L.A. sports world and our somewhat shaky beginning six months ago—losing our first home game in front of a half-filled Sports Arena—we've experienced two major sea changes, one on the team and the other in the stands.

Jerry, after his own shaky beginning, has blossomed as a player. The two of us have become, in my opinion, the most lethal one-two punch in the league. A writer calls us "Mr. Inside and Mr. Outside." Catchy, but we don't agree.

I like to drive the baseline or come off one of the wings and break to the rim, but I can hit from the outside as well. Jerry, while a knockdown shooter from distance, loves to drive to the basket. We both like to mix it up inside, we both love to rebound. Neither of us wants a nickname, especially one that limits us. Eventually someone dubs Jerry "Mr. Clutch." I think he accepts that one.

The second sea change involves the crowds. Ever since I scored 71 points against the Knicks and followed that with a six-game scoring flurry, I've noticed that the stands have been filling up more. By the time we begin our series against the Pistons, we're selling out.

"It's because of you," a few people tell me.

"It's not," I say. "We're playing well, a fun brand of basketball—"

"Elgin, it's because of *you*," someone else insists.

Well, I'll admit, I have gotten my share of press. Jim Murray, the revered *Los Angeles Times* writer, has started covering the team and written several flattering columns about me. We've also gotten covered by the "other" L.A. paper, the *Los Angeles Herald Examiner*. *Time* magazine and

Sports Illustrated write about everything from how I changed the game to my ball-handling skills.

Of course these articles flatter me. I'm human. Who wouldn't be flattered? But mostly I translate this praise into a validation of my success. I have succeeded in my chosen career. I will be able to feed and clothe my family, live in a comfortable house, buy some nice clothes for myself, and add to my record collection. But stepping back and thinking about all the attention, I ask myself, *Do I really deserve it?*

I know I'm a good player, maybe one of the best in the league, and I definitely take pride in how I play. I certainly feel blessed to be able to make a living doing what I love. Not many people can say that, especially people of color. But once I lace up my sneakers and make my first move or hit my first shot, everything else — all the noise, every distraction — flitters away and I go to work.

• • •

We start the best-of-five Western Division Semifinals with two games at home against the Detroit Pistons. I feel confident. We owned them during the regular season, and I can't see why that wouldn't carry through into the playoffs. We take them apart in front of two packed houses at the Sports Arena, 120–102 in the first game, in which I score 40, and 127–118, in which I drop in 49. We fly to Detroit for the next two games and I no longer feel confident. I've crossed over into cocky.

The Pistons ambush us in game three. We lose 124–113. We play as if every one of us has a piano lashed to his back. We don't look sluggish, we look sick. I sleep not at all. Three words bang through my head.

One more game. That's all we need to move on. One more win.

We don't get it in Detroit. In game four, I tear it up, scoring from every spot on the floor. I spot up and hit jump shots, cross over and drive to the hoop, slither, fake, and use the backboard at close range to knock in short jumpers. At one point, I shimmy past the guy guarding me and face Bob Ferry, the slick Detroit center, who stands between me and the rim like a six-foot-eight spike. I leap and dunk over him, causing him to cover his head. But Detroit smothers the other guys, and even though I score 47, we lose 123–114 and go back to L.A. with the series knotted at 2–2.

At least we're going home.

Home. For the first time, I attach that word to Los Angeles.

I grew up in D.C. I found a temporary home in Seattle, but I had to move on. I know the clichés. Home is where your heart is. Home is the place where, when you go there, they have to take you in. To me, home is where you feel the most wanted, where you belong.

Driving from the airport to my house, I press my face against the cool window of the taxicab, shut out Hot Rod's jabbering, and take in the L.A. I've come to know. The softly swaying palm trees. The lush green lawns in March. The boxy grid of the city spreading for miles below as we pass several sullen oil derricks.

If basketball's a city game . . . *the* city game . . . I would call this an odd city. But this city has taken me in. I belong here. I'm home.

The next night, playing at home in front of a screaming capacity crowd and with our new radio play-by-play announcer, Francis Dayle "Chick" Hearn — a local radio and TV guy whom Bob Short has hired for the playoffs, who's colorful, funny, opinionated, and golden-voiced — we trounce Detroit 137–120. I score 35, Jerry knocks in 25, and everyone else contributes.

Now we once again take on the St. Louis Hawks in the Western Division Finals, in a repeat of last year's bloodbath.

· · ·

We run.

Jerry, Hot Rod, Tommy, me — doesn't matter which of us handles the ball.

We *run.*

We sprint down the court.

The twenty-four-second shot clock seems too long. We'll take a shot in ten seconds if we can. Every possession—every make, every miss—leads to a fast break. Chick Hearn calls this the fastest game on earth.

We race like a track team dribbling a ball, whipping behind-the-back passes, draining stop-on-a-dime jump shots, and, when I'm the finisher, slamming down high-flying dunks above the rim. Twenty-plus years from now, Chick will label Magic Johnson's Lakers "Showtime," but we're beating them to it.

We're ahead of the pack and ahead of our time. I claim—especially with Jerry and me—that we are the original Showtime. Showtime 1.0.

The Hawks have had our number all year long. They play a slow, deliberate, physical game, even with their smooth, quick rookie point guard, Lenny Wilkins, who can run with the best of them. We still run, but we will have to cut through their forest of forwards—Pettit, Lovellette, Hagan, and Si Green, a leaper who's starting to come into his own.

First game in St. Louis, we trade baskets until I sink a jumper late in the fourth quarter and we hold on and shock them 122–118. I hang 44 on them, and Jerry puts in 27. The next night I score 35 points, but it feels as if they've got six or seven Hawks out there guarding us instead of five, and they swamp us by 15. Still, I'll take a split in St. Louis.

The crowd fills in early for game three at the Sports Arena and brings with it a buzz of excitement and anticipation. I know that many think of L.A. as laid-back. Well, not tonight. Not at the Sports Arena. These folks come in wired and loud and stay that way.

Ray Felix wins the tip, I score on a driving layup, and we claw to a one-point lead at the end of the first quarter. We pull away in the second quarter, take an eight-point lead at the half, and don't let up. We win 118–112, with five of us in double figures, taking a 2–1 series advantage.

Game four is a death match. Neither team can keep a lead. We never go up or down by more than four points the entire game. Jerry goes wild from outside, scoring 33, and I add 31, but we can't stop Pettit, who pours in 40. The Hawks squeak out a 118–117 victory that hushes the fans and ties the series at 2–2.

That game, even though it's a loss, seems to energize the city. Chick Hearn, the man of a thousand crazy descriptions and expressions (when a no-look pass I whip to Jerry sails behind him and out of bounds, Chick says, "The mustard's off the hot dog!"), insists we put on the best show in town. I appreciate that, and I begin to feel an actual connection to our fans. Losing in front of them leaves a knot in my stomach. I don't say a word to anyone, but I want to clock these Hawks in St. Louis and then come back and win the series at the Sports Arena—for our fans. I make that my mission.

We beat the Hawks badly in St. Louis, 121–112, as I score 47 and refuse to let them even get close. If Pettit or Hagan scores a basket, I don't match them—I make sure I score two.

Up three games to two, we go back to L.A. to finish what we started: dispatching the Hawks into oblivion.

The fans here at the Sports Arena tonight, numbering close to fifteen thousand, seem even wilder and rowdier than the previous packed houses. It's our night. For all of us, for the entire city. Everyone feels it.

Too bad we play like we've never seen a basketball before.

We can do nothing in the first half. Everybody on the team plays as if we've got board hands. We fumble easy passes, watching them skitter out of bounds. Rebounds slip through our fingertips. We can't seem to grip the basketball. We go down by 15 at halftime. We're lucky we don't trail by 30.

In the locker room, Fred looks as if his head is about to explode. Instead of bellowing like some kind of enraged beast, he clamps his mouth closed and barely speaks. He doesn't have to say anything—we all know how badly we played.

"Well," Fred says finally, philosophically, "you can't do any *worse*."

This deflates the tension. A few of us laugh.

"So in other words," I say, "your halftime adjustment is 'Play better.'"

More laughs.

"Hey, it's a new game," Hot Rod says. "Let's play like we're tied up, nothing–nothing. We can get back into this."

Somehow we do.

We fight back. We play the Hawks even in the third quarter and then we catch fire in the fourth. Everything happens in a flurry.

I slide away from the guy I'm guarding, block a Hagan shot from behind, and tip the ball to Jerry, who streaks downcourt for a layup. A few minutes later, Jerry steals the ball at midcourt and hits a long jump shot. I hit a turnaround jumper over Pettit. A Pettit shot clanks off the rim, I snatch the rebound, dribble the length of the court, spin right, and hit a three-foot running one-hander. Tommy Hawkins gets hot. Selvy comes alive. Rudy bangs the boards, tips in a Hot Rod miss. Miraculously, with six seconds left, we have erased the 15-point deficit and taken a 100–98 lead.

St. Louis ball out at midcourt. From the sideline, Fred barks out instructions, screaming through his cupped hands. "Let them inbound the ball into the backcourt, and then *foul* whoever catches the ball."

That's the only play. This way, because the infraction will occur in the backcourt, the Hawks player we foul will shoot only one free throw. We will get the ball back with five seconds left and a one-point lead . . . and they will have to foul us. Pretty much guarantees the win.

Every fan in the Sports Arena stands and screams. I've never heard the place this loud. The walls and floor vibrate. I smile. I love it.

We go into our defense and pick up our men. I can't make out faces. I see only uniforms. The ball comes into the guy I'm guarding. I wrap my arms around him and practically tackle him.

The refs don't call a foul. I keep pummeling the guy. *How do the refs not see this?* I've got my arms wrapped around the guy.

Somehow he gets the ball to Lenny Wilkins, who drives toward the hoop. One of our guys whacks Lenny hard, preventing the layup.

The refs call *this* foul—a shooting foul, awarding Lenny two free throws. He sinks them both. Ties the game 100–100.

We go to overtime.

Fred goes insane. He catapults himself off the bench and charges toward the closest official. Our trainer, Frank "Hoggy" O'Neill, a stocky wall of a man, throws himself in front of Fred and body-blocks him.

When Fred tries to bust past him, Hoggy throws his arms around him and restrains him from charging onto the court.

In the timeout before the overtime, Fred can barely speak.

"Calm down, Fred," I tell him. "Don't let the refs beat us. We'll get them."

We take the lead halfway through the overtime period, 107–104. On the next possession, Pettit drives to the hoop. He leaps. I leap higher and slap the ball out of his hands.

The official blows his whistle. He points to me, calling me for a foul. Number six. I've fouled out.

I bend over and do a Fred stomp. I start to protest to the ref, but I pivot away, slam my fists together, bang them against my cheeks, and run to the bench. The roar of the crowd cascades all around me, booming like a thunderclap. I sit on the bench and lower my head, the crowd noise growing, impossibly, louder.

Two minutes later, we lose the game by one point, 114–113.

Afterward, in the locker room, surrounded by reporters, Fred flips out, shouting about the officials, claiming that one referee in particular stole the game from us. He tells the reporters, who are scribbling furiously in their notepads, that he doesn't care what they write. He knows the league will fine him for disrespecting the referees. "Let them," he says, and then adds that in his seventeen years playing and coaching basketball, he's just experienced by far his worst loss.

It will be hard to come back from a loss like that.

We head to St. Louis for the seventh and deciding game of the series and give the Hawks all they can handle for forty-eight minutes. I score 39. Fred's tirade must've gotten to the refs because they hit me with only two fouls the entire game. I know I've committed at least two more fouls, but the refs swallow their whistles. Still, we lose by a basket, 105–103.

In the plane on the way back to Los Angeles, the team sits, darkly quiet. Finally, Hot Rod says, "Hey, we almost had 'em. We should've had them." He digs around in his pocket, pulls out a deck of cards, and turns to me. "Come on, baby, dealer's choice."

I try to distract myself, but the bitter taste stays, even as I deal seven-card stud. We fly home as the Western Division runners-up. The Hawks head to Boston to take on the Celtics for the NBA championship. They don't put up much of a fight. Boston crushes them in five games. I believe we helped Boston beat the Hawks, because we gutted St. Louis's spirit in our brutal seven-game series, a series we should have won. Small consolation.

I've had a great individual season. I averaged 34.8 points, 19.8 rebounds, and 5.1 assists per game, and I again make the All-NBA First Team, joining Wilt, Pettit, Bob Cousy, and Oscar Robertson.

I come home to L.A. planning to relax all summer, maybe take up tennis and golf, sports my teammates play in the offseason to help them relax. I like tennis and I see how you can become addicted to golf. I don't see at all how that sport can relax you.

Then, on a typically balmy Southern California day sometime in July, I sift through the day's mail. I come across a letter with a government seal. I slice it open with my fingernail. I read the letter and my stomach flips.

The letter is from the United States Army.

I have been called to report for active duty.

8

MEDIC

YOU CAN PLAY.

Tuesday, March 13, 1962.

I lie on my bed—well, my king-size cot—in the barracks at Fort Lewis Army Base, outside Tacoma, Washington. I have completed my second month of six months' active duty.

The way I see it, the army has discharged Elvis and called me to replace him.

I don't believe in conspiracy theories, but one night, lying in bed, mesmerized by the hum of the floor fan in the corner, it hits me: His name begins with *E-l*. My name begins with *E-l*. We both have five letters in our first names, and three of them—*E, l,* and *i*—are the same. We both make a good living *playing* in large arenas in front of thousands of screaming people. He's one of the top recording artists in the world, and I just signed a new five-year deal with the Lakers that makes me one of the highest-paid basketball players in the world. Coincidence? I think not.

I turn away from the fan and close my eyes.

I have way too much time on my hands. I'm going out of my mind.

At least neither of us has caused a war or an international incident.

I have four months of active duty left, so there's still time for me.

Some days, I still cannot believe I've been called up. But here I am, serving our country as global unrest heats up. Over the summer, the world saw East Germany, a Communist country aligned with Russia, construct a wall in Berlin, separating itself from West Germany, allied with us. In effect, East Germany and Russia drew a line in the sand in the form of this Berlin Wall, causing a worldwide case of the jitters. To prepare for a worst-case scenario, the U.S. Army has called up thousands of reservists.

So here I am.

Two months in and still nobody can find my uniform. I thought the army kept it. Maybe I misplaced it. Either way, I am once again without a uniform that fits. In the meantime, since I have been trained as a medic, I wear a green scrubs top — I think it's green — over my civilian pants and an army red cross armband pulled over my sleeve.

I care for victims of training accidents, soldiers who pass out on the obstacle course, soldiers who accidentally get shot on the firing range, and soldiers who come down with food poisoning in the mess hall.

Sort of.

I'm a medic.

I'm supposed to be learning to care for all these people, but I have two slight problems: I can't stand the sight of blood. And people who get sick actually make me sick.

So I assist, and try not to pass out as I provide bottles of water. Dab foreheads with damp cloths. Order supplies. Unpack boxes of supplies and put them onto shelves and into cabinets. File. Alphabetize. Schedule appointments. Cancel appointments. That's pretty much it.

Otherwise, when I'm not shooting baskets on the cruddy hoop in the tiny gym on the base, I lie on my bed, stare at the ceiling, and wait for word from my commanding officer that I can play in the playoffs, which begin in ten days: March 24 in L.A., against the Detroit Pistons.

When last season ended, I went through a day or so of feeling down, and then I perked up. I thought about our team — Jerry, a rookie, who really started to come on; Tommy, still young, getting better with each game; Rudy, strong rebounder, reliable scorer, only in his second or third year; Hot Rod and Frank Selvy, good players in their prime; and a solid bench backing us up — and I realized that we're just getting started. Then, a week after the season ended, Fred Schaus called to inform me that he had named me captain. I considered that an honor, and I told him so. Then, at the end of the summer, the bubble burst. The letter.

The letter told me to report to Fort Lewis in November. I drove to the Sports Arena, charged into Fred's office, and waved the letter at him. "The captain just got demoted to private."

He read the letter, cursed, and called Bob Short, who cursed and called Maurice Podoloff, the NBA president. Then Bob Short talked to someone with clout in the government and arranged to postpone my report date until January 2, 1962, allowing me to play more than half the season.

Fred still fumed. And he worried. He worried about playing the rest of the year without me. He worried about our growing and excited fan base staying home while I was away. He worried that the Lakers would make the playoffs and then get blown out. Fred grumbled to the press that we wouldn't be the same team without me.

And now I await permission from my commanding officer to participate in the playoffs.

I've played with a fury, the date January 2, 1962 — the day I would report to Fort Lewis — lit up in the back of my mind like a neon sign. We charged out of the gate, blasting through October and the end of November with a record of 15–3. Riding a four-game winning streak, we arrived in Philadelphia to take on Wilt and the Warriors, who, despite Wilt's averaging 50 points a game, were off to a mediocre start. Before the tip, Wilt sauntered over to me. He stopped suddenly and snapped off a crisp salute. "Thank you for your service, Private Baylor."

"You know, you can always enlist."

"Not my style," Wilt said.

"You're too big for the army."

"Well, I'm sure you'll miss me, but tonight I'm gonna give you a game you won't forget."

The game did become one for the ages. I scored 63 and we won our fifth game in a row, 151–147 in three overtimes. Wilt scored 78, pulverizing anybody who came near him. Afterward he sought me out, his hand extended.

"Shake my hand," he said. "I broke your record."

I looked down at his hand, hesitated, and reluctantly shook it. "Congratulations."

"Seventy-eight points. Told you I'd give you a game you'd never forget."

"If I'm not mistaken, we won the game," I said. "You're aware of that, right?"

"Watch your ass in the military," he said.

With only a few games left before my report date, the press converged on the team at the Sports Arena and asked some of the Lakers how they thought they would do without me. Rudy, speaking for everybody, said that the team would miss me and that this was an understatement. Hot Rod said he'd been working out the math, and he figured that if the Lakers could play .500 ball for the rest of the year and get me back for the playoffs, we'd have a real chance to win the championship.

"Just win every other game," he said. "That's all we have to do."

The day before I left for Fort Lewis, New Year's Day, we played to a full Sports Arena, again going up against Philadelphia. When the public address announcer introduced the starting lineups and called my name, everybody in the arena stood and applauded. We played a tense, tight game, finally pulling out a 114–111 victory as Wilt missed what seemed like 25 free throws and Jerry and I each scored more than 30. As I started to walk off the court, the crowd again stood, and this time they cheered. I stopped and waved. The cheering built, swelled. I kept my hand raised, mouthed "Thank you," and jogged into the dressing room.

• • •

Those three words. That's all I want to hear.

You can play.

The commanding officer, a colonel, holds my basketball-playing life in his hands.

The day after I reported to Fort Lewis, January 3, Fred Schaus called an emergency meeting of the owners.

"We've got thirty-eight regular season games left," he said. "If we can get the army to agree to let Elgin play on weekends, he can still play about fifteen games."

He expected wild enthusiasm. To his shock, he received a mixed response. A few of the owners didn't want me to play. They liked having me stashed away in upstate Washington in an army barracks, wearing army khaki, instead of in Southern California, wearing a Lakers uniform.

"We got a better chance of beating you without Baylor," one owner said.

"You're right," Fred said. "Good chance you lose if Elgin plays. *Excellent* chance you lose at the box office if he doesn't. Take your choice."

Silence.

"While you think about it, let me tell you that when we played Boston in the Garden at the end of December, we turned away three thousand people."

In the end, the owners sided with Schaus. They agreed to allow me to become a "weekend warrior," playing games on weekend passes.

Now the army had to agree, specifically my commanding officer.

The colonel informed me of his decision in person, fixing me with a steely stare. "I'll allow you off the base during certain agreed-upon time frames, with the following travel restriction: you cannot travel to a point that exceeds a distance of one thousand miles."

I blinked, trying to do the math in my head. The colonel read my mind.

"You're probably wondering, how far is Los Angeles from Fort Lewis?"

"Yes, sir, I was—"

"One thousand, one hundred six miles."

"That's—"

"Too far," the colonel said, his expression turning cold. "Yeah. Just a tiny bit too far."

• • •

After dinner one evening, I wandered around the base, trying to figure out what to do. I came up empty. I felt trapped. I returned to the barracks, deciding to turn in for the night. As I approached the barracks door, I saw a familiar figure leaning against the side of the building. I came closer and recognized Dr. Walter Brown, a prominent plastic surgeon from Seattle and a huge Seattle University basketball fan.

"Elgin," Dr. Brown said, stepping toward me and pumping my hand.

"Dr. Brown, what are you doing here?"

"Just checking in on a couple people I know. Figured I'd stop by and see you. I've been following your career. I'm proud of you. Had a nice season."

"Thank you. Yeah, we did all right. We can do better. We will do better."

"How you doing up here?"

I hesitated, debating whether to tell Dr. Brown how I really felt. I spoke before I could stop myself. "Not so good."

And then I spewed, telling him how the league was allowing me to play on weekends but the army wouldn't allow me to travel more than a thousand miles away from the base.

"It's frustrating," I said.

"Let me see what I can do," Dr. Brown said quietly.

"I don't want you to go to any trouble—"

"No trouble at all. You should be allowed to play. This colonel, what's this pipsqueak's name?"

I told him. Dr. Brown nodded. "We may have to go over his head."

I knew what he was talking about. I knew that back in Seattle, Dr. Brown once performed a procedure on a particular general's wife and that both the wife and the general were deliriously happy with the result. I also knew that the general outranked the colonel. Only a few days after I met Dr. Brown, I walked into the dining hall for breakfast. As soon as I entered the cafeteria, every enlisted man in the room stood and applauded.

"Thank you," I said, "but what did I do?"

"You got the travel restriction extended from a thousand miles to two thousand," someone shouted, and then everybody cheered.

Later the colonel visited me, wearing a grim robotic expression. "Clerical error," he said. "I meant to say two thousand miles."

"I'm sure you did, sir."

"And if you haven't heard, you've been granted special leave to play in the All-Star Game."

"That's great—"

The colonel cut me off. "You're a lucky man, Private Baylor."

He kicked his heels together. I snapped off a salute. The colonel swiveled on his squeaky heels and left.

"I don't believe in luck," I said after he was gone.

. . .

Russell. Wilt. West. Cousy. Pettit. Oscar. And me.

We all convened at Kiel Auditorium, in St. Louis, for the 1962 NBA All-Star Game. I assumed that every other All-Star was flying first-class round-trip, courtesy of their teams' owners. I bused from Fort Lewis to Seattle, flew commercial coach to Chicago, and then connected on a white-knuckle flight on a small plane to St. Louis. Delayed in Chicago, I nearly missed the tip-off.

I forgot all this when I stepped onto the court in front of fifteen thousand screaming fans. I hugged Jerry West and Frank Selvy, the Lakers' other two All-Stars, and shook hands with Fred Schaus, the West All-Stars coach. I hadn't played a game in two weeks, but I was sure once I got into the flow, I'd shake the rust right off. I hoped.

First possession of the game, Oscar darted between two East players, nabbed a rebound, and raced downcourt, dribbling ferociously, his head high. I sprinted ahead of him, burning down the left side. Oscar hit me with a perfect bounce pass and I knocked in a layup for our first basket.

Man, that feels good.

Oscar. Fluid, strong, intense, at times unstoppable. I respected his game as much as anyone's and later would support him as the leader of our players' union. I pointed my finger at him, acknowledging the pass, and then we got to work.

The East squad, with both Wilt and Russell, was bigger, but our West team, with Oscar, Jerry, Pettit, and me, was quicker and more explosive. We blew them out, 150–130. I scored 32, Oscar put in 26, and Pettit, with 25 points and 27 rebounds, won game MVP.

My favorite moment came late in the game. With my back to the basket, I posted up Willie Naulls, who checked me, bumping and shoving me. In my peripheral vision, I saw Wilt cheating over to double-team me, or at least provide Willie with insurance in case I slipped by him.

I backed into Willie with my butt, knocking him off stride, and then I faked left. Willie didn't bite. I leaned right. I could sense both Willie and Wilt poised to pounce, figuring that I would turn and try a scoop shot off the glass. But I didn't. I flipped the ball over my right ear . . . over Willie's head . . . over Wilt's head—

And banked in a no-look, over-my-*head* shot.

The crowd went nuts.

Willie grinned and shook his head. Wilt sniffed, ignored me, and hustled downcourt faster than I'd ever seen him run. Looked like someone fleeing a crime scene.

Some fifty years later, I would stumble upon a video of the top ten plays from that 1962 All-Star Game. That shot is ranked number one. The announcer never says who made the shot, never mentions my name.

• • •

In the next two and a half months, I played exactly six NBA games.

I played three games in Los Angeles—January 24, 25, and 28. I played February 18 and 25 in L.A. And I played March 12, when the Lakers flew to Seattle. That's it.

I never understood why the army, the team owners, and the league decided to limit me. I heard that some owners felt it was unfair if I was allowed to play against their teams and not against others.

I didn't know whether to make a legal stink and claim collusion, or just keep quiet and feel flattered by what I saw as the owners' respect. I didn't like either choice. I just wanted to *play*. And so I held my tongue, did my job, and counted down the days.

When I got word that I could fly down to Los Angeles on a Wednesday for a back-to-back against the Cincinnati Royals and then stay on for a Sunday afternoon game against the expansion Chicago Packers, I nearly screamed in the barracks. I tossed and turned all night, picturing myself on the basketball court, hoping that I hadn't lost anything in my two weeks of idleness since the All-Star Game.

Incredibly, my legs felt fresh and we beat the Royals on Wednesday night, 136–123. I scored 38 and Jerry caught fire for 50. After the game, I felt high. I couldn't wait for the second game, the next night.

I came out early, just to run a little bit, took a few extra shots, and shook off the cobwebs. I found my groove in this game, pouring in 39, and we beat Oscar and the Royals again, this time in a closer game, 116–112. I got a couple of days at home, in which I mainly slept, and then we took on Chicago, pounding them 124–109. Everyone pounds the Packers.

After the game, I said goodbye to my teammates, kissed Ruby and Alan, and hustled back up to Tacoma, checking into Fort Lewis before my curfew.

I didn't touch a basketball again for nearly a month. When I flew back to L.A. to take on Boston, I was so thrilled to be playing again that the Celtics didn't know what hit them. I played out of my skull, scoring 38, and we blew out Boston 125–99. Their entire team looked dazed. Red Auerbach glowered at me from the bench, his eyes narrowed, his cigarless lips pursed. I couldn't tell if he was astonished at how well I was playing after a month stashed away on an army base or whether he was disgusted with how poorly his team was playing. Maybe both. One thing about Red: you can't read him. He wears a poker face at all times.

A week later, I flew back down to L.A. for another Sunday afternoon game, this time against the Detroit Pistons. We crushed them, 128–99, as I scored 45. Finally, two weeks later, on March 12, the regular season coming to an end, I joined the Lakers in Seattle for a game against the Knicks. We beat them easily, 119–106.

We've won all six games I've played while stationed at Fort Lewis, and I've averaged 39 points a game. We finish the season with a 54–26 record, eleven games ahead of the second-place team. In the forty-eight games I played, we've gone 37–11. With our record, we've earned a first-round bye in the playoffs. On March 24, we will take on either the Pistons or the Royals in a seven-game Western Division Finals series, the winner going to the NBA Finals.

And now I lie on my cot in the barracks, the blades of the floor fan clacking in the corner. I listen to the sound of my own breathing. The minutes tick by. My life — at least my immediate future — will be determined by the colonel and three simple words.

On March 21, the colonel enters the barracks and hands me a letter on official army stationery. I try to read it, but my hands start to tremble and the words swim in front of me on the page, morphing into an indecipherable inky blot. The colonel clears his throat. He speaks quietly.

"The letter says the army will grant you leave, but you must report back to Fort Lewis on April 19. No later."

"You mean —?"

"Yes," the colonel says, and then speaks the three words I've been waiting weeks to hear.

You can play.

• • •

I can't explain it. I should feel out of shape, winded in the fourth quarter. I should feel tight, my timing off, my muscles slack. I should need some extra rest, a blow here and there. I should need a game or two, at least, to get my timing back. I've missed thirty-two games. I haven't played regularly in two and a half months.

But I feel loose, strong, and ready. So ready.

We open the Western Division Finals at home against Detroit. We rout the Pistons in the first game, 132–108, as I score 35, and then we trounce them 127–112 in the second game as I score 40. I know they'll play us tough in Detroit, but I remind everyone that I have to be back at Fort Lewis on April 19, so we can't waste any time. In game three, we blow a big first-quarter lead but hang on to win 111–106. We go up three games to none. I want to put this series to bed. I want to sweep them.

Doesn't happen.

Game four. The Pistons take the lead. We fight back. I score 45, Jerry puts in 41, but we can't catch them. The Pistons cling to a one-point lead and we can't get past them. We fall 118–117. I'm confident that we'll knock them out in game five in L.A.

Doesn't happen.

In game five, in front of a sold-out, crazed crowd in the Sports Arena, we play like crap for three quarters.

Maybe my playing so sporadically for two and a half months has finally caught up with me. Maybe my legs feel heavy. Maybe I'm winded. Or maybe we just stink.

As we begin the fourth quarter, we find ourselves down by 28 points.

I don't say a word. Fred has set a new record for stomping. He's yanked off his suit jacket, balled it up, and tossed it near the scorer's table. He's gone so red in the face that I'm afraid he's about to stroke out. I take a deep breath as we form a huddle around Fred. I feel as much to blame as anyone for stinking up the Sports Arena.

"Hey, guys," I say, peeking up at the scoreboard. "We're down by 28. Let's have some fun. See if we can make this a game."

I score three baskets in a row, Jerry makes a steal, and we're off. We knock the Pistons backward. We close the gap to 20 points, then 18, then 15, then 12. Detroit coach Dick McGuire calls a timeout, and I grin. The Pistons have stopped playing us; they're playing the clock. We may not beat them tonight, but we have changed the momentum of the whole series. They won't beat us again.

Our comeback falls short. We outscore them 51–30 in the fourth quarter, but they hold on and win game five, 132–125.

We take the series in Detroit, winning the fourth game of the seven-game set, 123–117. Jerry and I each score 38, and while the score makes the game seem close, we never trail.

Now, at last, we will play the Celtics for the NBA championship. I can't explain the intensity I feel worming through my stomach. I don't feel nerves. I feel . . . desire. I want this. I want to beat them so badly.

In the bowels of Boston Garden, we change in our freezer of a locker room, with one working toilet and a shower that sporadically trickles ice water. I bring in a portable record player to keep us loose, playing 45-rpm singles like "Duke of Earl," by Gene Chandler, singing along and getting guys like Krebs and Rudy to join me. We feel loose. We laugh about the locker room conditions, joking with one another and with reporters. A few guys puff pregame cigarettes. I bum a smoke off sportswriter Frank Deford, take a few drags, and blow the smoke into the ceiling vent that pumps out frigid air. Funny how the best reporters know what they can write and what they know to be off the record. We never need to admonish them. They just know. I've learned to respect and even trust writers like Jim Murray and Deford, because I know they won't burn me. The reporters who cover the team, along with Chick Hearn, travel with us and become part of our road entourage.

I become especially close with Chick. On plane flights, we start what becomes a never-ending two-man gin rummy tournament. We talk basketball, we talk life, we confide in each other. Once, as I'm walking through Logan Airport after a game against the Celtics, a sudden sharp pain tears through my abdomen. I grip my midsection and collapse onto the floor of the terminal. I wake up the next morning in a hospital bed with a stomach ailment, treatable with medication. Chick sits curled up in a chair next to the bed, conked out. He has spent the night sleeping in that chair. He stays with me until the doctor releases me from the hospital the following day.

Now, in our locker room in Boston Garden, I stamp out my cigarette, step over the bench, and sit down next to Jerry.

"You carried us when I was gone," I say.

"I tried," Jerry says. "We can beat these guys, but not without you." He shoots me a thin smile. "Welcome back."

Game one. We come out strong, take a lead in the first quarter, but stumble and fade in the second. The Celtics come at us in waves, a different guy scoring every basket. Seven Celtics put up double figures and they beat us 122–108. I score 35, but I walk off the court dejected. I walk past the reporters, shaking my head slowly. I don't feel like talking. The anxiety squirrels back into my stomach.

We play a different game in game two. Jerry finds a new gear — overdrive. He shoots from all over the floor and gives the Celtics' guards fits as they try in vain to chase him. He pumps in 40, I score 36, and we win 129–122. We head back to Los Angeles with the series tied at one game each.

A record crowd packs the Sports Arena for game three. Not only have we begun to outplay the Celtics in this series, but we've outdrawn them. Warming up before game two in Boston, I was shocked to see empty seats. Here at home, I see people standing two deep along the back wall. And in what will become a Lakers tradition, celebrities sit courtside — movie and TV stars I recognize, like Pat Boone, Doris Day, Danny Thomas, and Dinah Shore.

Game three. I find my stroke early, managing to spin away from Tom "Satch" Sanders, who's guarding me, and hit several long-range jump shots. Jerry heats up, too, and, going into the fourth quarter, we lead the Celtics by 12.

Then I miss a couple of shots, Jerry goes cold, Russell takes over the middle, swatting shot after shot, and the Celtics chip away at our lead, finally catching us, 115–115. With four seconds left, we miss the go-ahead shot. Russell snags the rebound and calls timeout.

The Celtics take the ball in beneath their basket. I know Red has drawn up a play intending to get the ball to Sam Jones or Bob Cousy for a long desperation shot from just inside half-court. I say something about this to Jerry, suggesting that we play those two guys closely while watching Russell's eyes as he inbounds the ball.

Russell takes the ball out. He snaps a short pass to Sam Jones, who fires a long pass for Cousy.

Jerry reads the play perfectly. He darts in front of Cousy like a defensive back covering a wide receiver and picks off the pass. Without breaking stride, he glances at the clock.

Three seconds.

Not enough time for him to get to the rim.

Everybody on our bench leaps up. Somebody shouts, "Shoot!"

Jerry streaks toward the hoop.

Two seconds. He can't make it. He doesn't have enough time.

Shoot!

He flies by the last defender. Russell.

SHOOT!

He keeps going . . .

One second. Jerry lays the ball into the basket and time expires.

We win, 117–115.

The place explodes.

I look at Red. I can't be sure, but, reading his lips, I think he says, "What the fuck?"

Fans spill out from the stands and swamp Jerry. I wave, try to catch his eye, but he can't see me. He extricates himself from the scrum of people and dashes into the dressing room.

We swarm him there, pummel him.

"I saw the clock," he says. "I had three seconds. I knew I had enough time."

That's what he says aloud. But I see more in his eyes.

Elation. His eyes say *elation*.

The next night, leading the Celtics two games to one, the night after Jerry's steal of Sam's pass and his miracle layup at the buzzer, we lose to Boston 115–103. I score 38, but a bunch of Celtics score 10 or more and they tie the series. I so want to win this one for our fans who've once again filled the Sports Arena, but we fall behind in the first quarter and never really make a game of it. We will now have to win at least one game in Boston to take the championship.

·   ·   ·

April 14, 1962. Game five of the NBA Finals.

I take my position for the opening tip. I shake my wrists to loosen up. Pump my legs. I glance at Bill, who turns his head and ignores me. The sold-out Boston Garden booms behind me.

I decide I want to come out shooting to try to give us an early lift. I fear that if we fall behind, we'll dig ourselves a hole like we did in the last game and we won't be able to climb back out. So from the opening tip, I look for my shot.

I keep shooting. I mix up my shots, pop from outside when I've got a good look, drive to the hoop when I see an opening.

At the end of the first quarter, we take a one-point lead.

I face double teams all night, but for the most part Satch Sanders, the Celtics' best defender, checks me. He likes to stab me in the midsection or ribs with his fingernails, which he keeps sharp as blades, trash-talking nonstop. I tune him out. And while I can't stand his jabbing me in the gut, I use it to my advantage: when he pokes and prods, I know exactly where he is.

I feel his fingers and watch his feet. If I get him to bite on a shoulder or head fake, he'll cross his feet, I'll tie him up, and he becomes vulnerable to my spin move and my step-back jump shot. And so I do. I drill a bunch of jumpers over him, and when he bodies me up, I shake once and drive past him.

We keep the game close at the half and into the third quarter, and then the Celtics go up by six to begin the fourth. By this point I have a vague sense that I've been scoring more than usual. I know for sure that my shot has been dropping—I don't recall many misses. Now, when Satch leans one way, I blow by the *other* way, pretty much at will, or cross over and speed by him the first way. Bill starts coming over to help, but I fake him out once, spinning by him and putting in a finger roll, and another time I take off at the right wing and soar, dunking over him so hard and so fast, he can only retreat.

We grab the lead midway through the fourth quarter. I don't want the Celtics making any sort of comeback, so I hit a couple of jumpers, a

floater, and a running one-hander. Insurance policies. We win 126–121
to take a 3–2 series lead, and when the game ends, to my surprise, the
Boston Garden crowd stands and cheers as I leave the court.

"Why are they applauding?" I ask Hot Rod. "Their team lost."

"True dat, baby," Hot Rod says. "You also scored 61 points."

"I did?"

"Yeah. No big thing. You just set the all-time record, that's all."

I honestly had no idea how many points I scored or that I've established the record for most points in an NBA Finals game.

Michael Jordan will eventually score 63 in an NBA Finals game, but
he'll need overtime to do it.

So I guess my record still stands.

. . .

We got them.

We're coming home up three games to two, in front of the most
frenzied, sold-out house yet. These fans, in only their second season,
smell blood. L.A. loves winners, and now they want an NBA title.

Game six. We go up by eight at halftime, and I feel it. I practically
bolt out of the locker room to begin the second half. We've got the
Celtics under our thumbs. Krebs and Ray Felix have kept Bill Russell
quiet and pushed him around enough under the boards that he can't pull
down his usual million rebounds. We've locked down Heinsohn, Cousy,
and the rest of them.

Except Sam Jones. Their sharpshooting guard goes off. We all try to
check him—Jerry, Hot Rod, Selvy, Tommy, and, finally, me.

The Celtics run him behind screens, sometimes double screens, and
Sam can't miss. Sam's a talker—not a trash talker, really, but when he
starts drilling in jumpers and I can't quite get to him through a Heinsohn
or Russell pick, Sam grins and says, "Too late."

In the end, Jerry and I each have 34, but Sam carries the entire Celtics team on his back. He scores 35 and they take game six, 119–105.

Back to Boston for the seventh and deciding game.

. . .

April 18, 1962.

After a choppy cross-country flight, I order room service and turn in early. I manage maybe four hours of sleep, none of them in a row. I can't stop thinking about this series, running moments through my mind, thinking what might have been, what should have been, if only . . .

Three words.

Three words bludgeoning my skull.

We had them.

I imagine myself at home in D.C., watching wrestling Friday nights with Pops, and suddenly I see myself in the match. I am Argentina Rocca, and I have Killer Kowalski flat on the canvas, helpless, beaten, pinned, and then somehow, according to the script that everybody knows, he twists out of my grasp and squirms away, waving his finger, taunting me.

We had them.

Ninety minutes before game time, I take a cab to the Garden with Hot Rod and Tommy. The cabdriver pulls over and stops a block away.

"Close as I can get," the cabbie says.

A crowd floods the street, thousands of people desperately trying to score tickets for game seven.

"People been here since three in the morning," the cabbie says.

Later I'll hear that the Boston Garden box office turned away seven thousand people.

The Garden this night feels electric. Red starts Satch Sanders on me early, then switches to Heinsohn and brings in burly Jim Loscutoff basically to bang me whenever I touch the ball. My shot doesn't fall at first, so I drive, stopping abruptly for short jump shots or gliding to the rim for a variety of scoop shots. On the Celtics' side, Russell pulverizes Krebs and Felix, scoring on them and completely controlling the boards. Jerry hits a few jump shots, and even though we clamp down on Sam Jones, the Celtics take a six-point lead at halftime.

We catch them at the end of the third quarter, and, with the energy in the Garden crackling, the din of the crowd migraine-inducing, the uneven parquet floor with its occasional dead spots rumbling, the score seesaws back and forth. Finally, Boston creeps ahead by six points. Late in the quarter, Heinsohn whacks me across the arm, committing his sixth foul. I drop in the two free throws to cut the lead to four points. We stop them on the next possession and Jerry hits a jumper to trim Boston's lead to two. Then the game tilts in a surprising direction. Frank Selvy comes to our rescue.

Frank, a tremendous shooter, has become a complementary player, usually our fourth or fifth option behind Jerry, me, Rudy, and Hot Rod. What people may not know is that when Selvy played at Furman University, he was a legendary scorer, one of the greatest college players ever. Frank once scored *100* points in a game. More than Wilt ever did in college. More than anybody. I'm never surprised when Frank, a modest, sensitive guy, makes two or more jumpers in a row. I'm surprised when he doesn't.

Down 100–96, Frank slithers in for a rare rebound, turns, and dribbles the length of the court for a layup. Moments later, after a Celtics miss, Selvy misses a jumper, but he follows his shot and tips in his own rebound to tie the game, 100–100.

Then the Celtics miss again and we get the ball back.

Ten seconds left. Hot Rod quickly brings the ball up and looks for me. I wave my hand, but Russell slides in front of me and blocks Hot Rod's vision. The clock ticking down, Hot Rod searches for Jerry. He can't find him. But he sees Frank, standing alone inside the left baseline. Hot Rod slings a pass to Selvy, who rises up for a twelve-foot jump shot, a gimme, to win the game. At the same time, I race to the hoop to get in position for an offensive rebound. I crouch and follow the flight of the ball as it arcs out of Selvy's fingertips and —

Clangs off the back rim.

I leap for the rebound to tip the ball in.

Someone shoves me in the back.

I tumble out of bounds.

Behind me, Bill Russell snatches the rebound as the horn sounds, ending regulation, the teams tied 100–100.

I whirl to find the official, who has his back to me. I rush over to him. "He pushed me . . ."

His whistle lodged in the corner of his mouth like a cheap cigar, the referee says, "I don't want to hear it, Elgin."

"But . . . I swear . . . he *pushed* . . ."

I turn around and see Sam Jones smiling at me. Sam raises his shoulders in the tiniest shrug.

Play on. That's all we can do.

But we do nothing in overtime. I miss two shots and then I foul out. We never come close in the extra period. We lose 110–107 as Bob Cousy dribbles out the clock. The Celtics repeat as NBA champions. I head toward the locker room, my head down. I scored 41, but Russell scored 30 and pulled down an astonishing 40 rebounds.

After the game, our dressing room feels like a tomb. We all sit frozen in front of our lockers, none of us able to speak. I keep seeing the two shots I missed late in the game. I feel Sam Jones's hands on my back, shoving me, and I wince. Hot Rod bounds in, his eyes wild, his face even redder than usual. He goes straight to Frank Selvy, who sits slumped, genuine agony lining his face.

"Hey, baby, don't worry about it," Hot Rod rasps to Selvy. "Get your head up. You only cost us about $30,000."

Ray Felix lays a gentle hand on Selvy's shoulders and says, in all sincerity, "Don't listen to him. We'll get 'em tomorrow."

Within a few minutes, I shower, change, and head out of the dressing room. I have had my best season. I've averaged 38.3 points a game — what will still stand, in 2017, as the second most in NBA history, behind Wilt's 50.4 per game this same season — in addition to 18.6 rebounds a game, and the league names me to the All-NBA First Team, along with Jerry, Bob Pettit, Oscar, and Wilt.

After the game seven loss, I walk slowly out of the players' entrance, my head lowered, my eyes half closed. I will meet Bill for dinner. We will not talk about the game, or at least I won't. I can't. We'll talk about the future. Tomorrow I will return to Fort Lewis to complete my six months of active duty. But right now all I can see is three new words flashing in front of me, tall as buildings:

We had them.

. . .

Summer 1962.

I leave Fort Lewis and come home to Los Angeles to prepare for the upcoming season.

At the beginning of the summer, I speak with Fred Schaus and Lou Mohs, our general manager, both of whom seem determined to do whatever it takes to bring an NBA championship to L.A. We all agree. It's clear that what we need in the draft is a center who might be able to neutralize Bill Russell.

Mohs doesn't draft a center.

He drafts *two*: six-foot-ten All-American Leroy Ellis from St. John's, a speedy, fiery scorer, who plays with a mean streak and likes to mix it up, and six-ten Gene Wiley, an extremely athletic shot blocker and defensive specialist out of Wichita State, who's one of the quietest, most sensitive people I will ever meet and is always sketching and painting in his ever-present sketch pad. Unfortunately, during their rookie seasons, Leroy fights more than he scores, and Gene paints better than he plays. But at least they provide reinforcements. With Jim Krebs, we can now throw three centers at Russell and Wilt, or, at the very least, we can foul them eighteen times a game. That should slow them down a little.

Lou also brings in Dick Barnett, a welcome "third option," a quick, streak-shooting guard he pries away from the Syracuse Nationals at a hefty price.

He also deals Tommy Hawkins to the Royals, probably to cut payroll. It's just business, I know, but I'm sad to see him go.

As I get in shape for the season by playing pickup games at the Pepperdine University gym, I start to hear the buzz about our basketball team. The Lakers seem stacked. We will dominate the West again and this time beat Boston. We have become the hottest sports attraction in town. I find myself featured in magazine articles. Apparently the face of the NBA, even more than Wilt or Oscar, I make the cover of *Sports Illustrated*. Gossip columnists talk about us, and movie stars make Lakers games a celebrity destination, *the* place to be seen. The league gets in on the action, wanting to take advantage of the heat. L.A. will host the 1963 All-Star Game at the Sports Arena. And this season, the Philadelphia Warriors will move to San Francisco, easing our travel schedule considerably. The league has also changed the name of the expansion Chicago Packers to the Chicago Zephyrs. The Chicago *Zephyrs?* Sweet, gentle breezes? Not sure that name works.

We start the season with five games on the road. We lose four of them. Dick Barnett fits right into the starting lineup, but Fred fools around with alternating Krebs, Ellis, and Wiley. Overall, we seem very sluggish. Actually, if I'm honest, *I* feel sluggish. I feel fine physically, but I can't seem to get my motor going. I'm not sure why. Could be that last season's two-and-a-half-month layoff has finally taken its toll. Could be that I'm still smarting from the brutal loss to the Celtics. Could be that I'm just fatigued.

As we head back to Los Angeles for our home opener, I decide that in order to get myself right, I need to play through it. Just put my head down and play. That's what I always do. When I hit a rough patch emotionally, from my adolescence through my hard times in high school, if I felt down or confused, I'd pick up a basketball, find a game, and play through it.

But now? I don't know. Maybe I'm feeling something else, something different, something new.

I'm in the prime of my career. I'm the toast of the town, the captain of the defending Western Division champs. We've reloaded over the summer, our sights set directly on our nemesis, the Celtics, the only thing that stands between us and our ultimate goal, a ring.

What if we don't get there? What if we fall short again? What if we have gone as far as we can go?

I'm playing as well as I can play, as well as I've ever played, and — if you believe the so-called experts and basketball gurus — as well as anyone has *ever* played. And what if it's not enough? What if I have reached my peak?

I don't feel unsettled or scared. I feel . . . vulnerable.

Thoughts snake into my head at night now, thoughts that keep me awake, disturb me. I manage to push them away, ignore them. Except for one. The one thought I can't seem to shake.

What happens when I reach the end?

How do I play through that?

• • •

I snap out of my funk. In front of a capacity crowd for our home opener, I score 31 and we blast a weak Detroit team, 134–118. Then we reel off five straight wins.

One night, playing the San Francisco Warriors as part of a four-game home stand, I notice that Wilt, looking surlier than usual, appears to be on a mission to score every time he touches the ball, which I interpret as some kind of direct message to me. Strange, because this night, while I score 30, Jerry goes off, scoring 49. But Wilt seems to be playing his own game within a game, oblivious of the score of the actual game, focused only on his point total, which reaches 72, even though we beat his team by 12. For extra sadistic pleasure, he seems intent on butchering all three of our centers.

During the third quarter, Fred has had enough of this and calls a timeout. As we huddle in front of our bench, Fred frantically scans our team and fixes his stare on — my memory fails me on the exact player, but we'll say Jim Krebs.

"Listen," he says to Krebs, and then lifts his head and looks over at the Warriors' bench. He catches Wilt's eye. Wilt glares back at him, at all of us. "Get out there and foul Wilt. Don't let him shoot."

"You want me to—"

"Yeah. *Foul* him. Hard."

"Are you sure—"

"Do it."

We break the huddle and resume play. As we bring the ball up, Krebs doesn't seem happy about his assignment, but Fred gestures wildly at him and points at Wilt. Krebs sighs, swallows, and starts to run toward Wilt.

Either Wilt heard Fred in the huddle or he can read lips, because as Krebs approaches, Wilt widens his legs into a fighter's stance, pulls back his fist, and says, "If you put your fucking hand on me, I will knock you out."

Krebs skids to a stop and then retreats. He jogs past Fred, spreads his hands helplessly, and hollers, "Coach, I can't foul him!"

"Why not?"

"He won't let me!"

• • •

We catch fire. We win six in a row, drop a couple, split the next two, then win nine in a row, then a short time later we win *eleven* in a row. Everyone contributes. Jerry and I take turns leading our team in scoring, both of us averaging more than 30 points a game, with Dick adding another 20 almost every night. Although it's too simplistic to say that Jerry shoots mainly from outside and I tend to score on driving layups and dunks or unorthodox, improvised scoop shots, we maintain our reputations as Mr. Inside and Mr. Outside. At one point, Fred suggests that I focus more on my outside shooting.

"May come a time," he says, "that you need to develop your outside shot. Not saying now. Not saying immediately. I'm saying in a couple of years, when you're, you know . . ." He bobs his head and allows his voice to trail off.

"What, Fred? When I'm what? Older? Slower? What's on your mind, Fred? You can say it."

"I'm just saying you should shoot from outside more, that's all."

I do start to shoot more from outside, not necessarily because Fred told me to, but because teams do sag off me to try to cut me off from driving, and that gives me an open outside shot. Fine with me: I'll take it. I also develop a knack for missing short shots on purpose. I do this so I can follow up my own shot and tip in my miss, playing the angle, a move that works well against teams with big, burly centers. Actually, this season, every team in the league seems to have a big guy in the middle. In addition to Russell and Wilt, we face Zelmo Beaty of the Hawks, Walt Bellamy of Chicago, and Wayne Embry of Cincinnati.

I believe that as a team, we mesh well on the court because we get along so well off the court, especially on road trips. We're different people, no doubt, coming from various backgrounds and parts of the country, black guys and white guys, but on the road we become a basketball band of brothers. It usually takes me time to warm up to people I don't know, but once I'm comfortable, I relax and let loose. It's not unlike figuring out who you are on the basketball court. I stay back a little, assess who you are, how you play, what I can get away with. Same thing socially: you can tell how comfortable I feel by how much I talk. I've come a long way from that shy sophomore at Seattle U. Now, especially when I'm relaxed — when I'm me — I talk. Not just occasionally. I talk all the time.

"He never stops talking," Rudy tells the guys one night at dinner. "I mean *never*. He falls asleep talking and he wakes up talking. He has no on-off switch."

After Krebs gives me the name Motormouth, I start giving other guys nicknames. As I said, I call Rudy "Musty" because he refuses to use deodorant and seldom showers. I call Gene Wiley "Gabby" because he never speaks. I call Hot Rod "Rodney" because he already has a nickname. I call Frank Selvy "Pops" because he shoots from outside and he looks old. I called former Laker Larry Foust "Board Hands" because he couldn't catch a pass. It was as if he had hands made of wood.

We joke with one another. We crack each other up. We make silly prank phone calls, ordering expensive room service meals and sending them to unsuspecting players. We make fun of people's taste in food, women, and especially clothes. Inspired in part by Dick Barnett, a clotheshorse, I upgrade my wardrobe. I hire a well-known Beverly Hills tailor to custom-make my suits and shirts, and I buy long, stylish overcoats. I wear muted colors — tan, camel, gray, and brown — so I don't have to deal with my color-blindness. I dress well because I enjoy looking sharp, but I also believe that NBA players have become public figures, borderline celebrities, and that we should take pride in how we look. More than once when I see how a rookie on our team dresses, especially guys from poor, usually rural areas, I take them clothes shopping. I know they don't make much money, so I spring for their new wardrobe.

"You don't have to do that," one guy says.

"Yes, I do," I say. "Those hick clothes you wear? I can't be seen with you."

For some reason, as a team, we go after poor Frank O'Neill, our trainer, mercilessly. We call Frank "Hoggy" or "Piggy" because of his uncanny resemblance to a certain well-known cartoon character. I think we also go after Hoggy because he's so serious, which makes him the perfect target. Once, during a game, I twist my ankle. I hobble over to the sideline, plop down on the bench, and wait while Hoggy expertly tapes me up.

"Great job, Hoggy," I say after he finishes. "But you taped the wrong one."

I stifle my laugh as long as I can, but Hot Rod, sitting next to me, can't contain himself. We both lose it.

"Funny," Hoggy says, not finding this funny at all.

From that moment on, whenever Frank tapes someone up, that guy always says, "Hey, Frank, nice job, but you taped the wrong one."

One time, in New York, we get Hoggy good. It takes days of planning and preparation.

Hoggy always enjoys a couple of cocktails in the evening, especially after a big meal or when someone else pays. After our game against the Knicks, a couple of guys take him to the hotel bar while I run out to a butcher shop around the corner and buy a pig's head. In the meantime, two other teammates commandeer an extra room key and go to work in Hoggy's hotel room. By the time I arrive with the pig's head wrapped in butcher paper, they've finished decorating. They've set up candles on the nightstands and arranged three small lamps on Hoggy's bed. I remove the pig's head from the butcher paper and place it in the center, between some pillows. I fluff the pillows and the guys turn on the lamps, which illuminate the pig's head perfectly. The bed looks like some kind of weird hog shrine.

We head down to the bar and have a drink with Hoggy, then we all escort him to his hotel room. He opens the door, enters his room, sees the pig's head on his pillow, and screams.

We crack up.

He starts to shake. For a second, I think he's about to have a heart attack, but then I realize that he's laughing.

"Funny," Hoggy says. "Very funny."

· · ·

I wouldn't say I'm obsessed with beating the Boston Celtics, but the thought does occupy my mind late at night and sometimes filters into my dreams.

A day or so before the 1963 All-Star Game, held in Los Angeles, a local sportswriter calls L.A. the center of the basketball world. Red Auerbach, coach of the East All-Stars, goes crazy. He reminds reporters at a luncheon that his Celtics have won four straight NBA titles and suggests that the Lakers should win a couple of championships before anybody considers L.A. the *center of the basketball world*. He says that Boston will remain the center for a very long time.

Bill Russell, I think.

He's all that keeps us from that claim, from making L.A. the center of the basketball world. From beating them.

I love Leroy and Gabby and Krebs, but all three of them together don't add up to Bill Russell.

We have firepower. We have Jerry and we've got me. No team has ever had two players who score the way we do. That's not bragging; that's fact.

And now we've got Dick, who can light it up, and Rudy . . .

But they've got Russell.

. . .

We go on an eight-game winning streak, with six of those eight wins coming on the road, and lift our record to a gaudy 43–12. We arrive in New York for the last game on the trip, taking on the weak Knicks, a team everybody beats. We start off slowly in the first quarter, fall behind by a few points, then pull it together in the second, though we still trail at halftime.

Early in the third quarter, reverberating through the Madison Square Garden crowd, I hear a horrific shout and what sounds almost like a gunshot. I turn and see Jerry on the floor, writhing and gripping his left leg. I run over to him and a couple of us help him off the court. Hoggy quickly examines his leg and shakes his head. Clearly, Jerry has sustained a serious injury. Hoggy guesses hamstring. Distracted, we play horribly the rest of the game and the lowly Knicks blow us out, 122–95.

We fly back to L.A. for two games against the Chicago Zephyrs before we hit the road again. Back in L.A., I learn that Hoggy called it: hamstring. Seven weeks' recovery, minimum, meaning we lose Jerry for the rest of the season. Hopefully he'll be back for the playoffs.

We play Chicago three straight games. We win all three, but barely, each game a squeaker, which I find concerning, since the Zephyrs own the second-worst record (as well as the worst name) in the league.

I look over our schedule. We have twenty-one games left in the regular season, nearly half of them against either Boston and Russell or San Francisco and Wilt. The Warriors aren't very good, but we can't match Wilt, and they always make us work. Without Jerry, I'll just have to step it up.

In the next six games, bringing us to the middle of February, I score

42, 35, 50, 41, 46, and 42 — an average of 42.7 points a game. After that we split two games with Boston, and then we play the Detroit Pistons in Des Moines, Iowa, and beat them, for our fiftieth win of the season, as I score 49.

Then something happens.

Something happens that at first I refuse to acknowledge and then I keep to myself.

My knees start to hurt.

They have been aching, from time to time, if I think about it, for months, maybe even years. I handle the pain in the best and most convenient way I know: I ignore it. Once in a while, teammates notice me grimacing and ask if I'm all right. I smile, shake my head, wave them off, or sometimes pretend I don't hear them.

But, heading into March, the pain starts to come on more often and more acutely, especially when we play Sunday afternoon following a Saturday night game. One Sunday afternoon against the Pistons, my knees feel stiff the whole game. I can usually run through the stiffness, loosen up my joints either after warm-ups or early in the first quarter, but this day I can't work out the stiffness. I score 21 painful points and we lose, dropping our seventh of eight games.

It's stress, I tell myself — physical and mental stress.

We end the season at home against San Francisco, a game that goes into overtime before we beat the Warriors, 111–105. Wilt scores what is for him a disappointing 40 points, and I drop in 37, but by the end of overtime, pain is shooting through both knees, especially my right knee, and I wince with each step. As we walk into the tunnel leading into the locker room after the game, Hoggy catches up to me. "You're hurting," he says. "I can see that. Anybody can."

"I'm fine, Hoggy. Long season."

"Tell that to them."

He nods toward two guys in ties who are waiting for me: our team doctors, Ernie Vandeweghe and Chuck Aronberg, who voluntarily attend every home game on their own time and their own dime. NBA teams employ trainers, guys like Hoggy — a magician with surgical tape and a wizard with a can of first-aid freeze spray—but they do not routinely hire full-time doctors. I respect Ernie and Chuck. I do: they're great guys and real doctors. I'm just always a little hesitant when they dispense medical advice, because Ernie is a pediatrician and Chuck is an ophthalmologist.

Chuck steps forward and walks alongside me as I step slowly into the locker room. "Great game, Elgin. Always love when you beat Wilt."

"Thanks. Me, too. Excuse me. I want to take a shower—"

"Listen, Elgin," Ernie says. "You've been doing a good job concealing the pain in your knees—"

I stop cold.

Chuck gently touches my arm. "We think you need to see Kerlan."

Robert Kerlan: sports medicine guru, orthopedic surgeon to the stars, the Los Angeles Dodgers' team doctor.

"Have you guys been talking to Hoggy?"

"We can see it," Chuck says.

No sense denying it.

"Okay, so, you two"—I search Chuck's face, then Ernie's—"think the knee pain means there's something serious?"

"We think you need to check it out."

"Call Kerlan for a consultation," Ernie says.

"I will," I say. "After the playoffs. I promise."

I keep my promise. I see Dr. Kerlan—a year later.

. . .

We begin the playoffs two weeks later, at home against the Hawks. I spend the long layoff doing nothing except attending practices and hanging around the house, resting my knees.

I think about our upcoming matchup with the Hawks. We've won the Western Division for the second straight year, finishing 53–27, but the Hawks were only five games behind us. They play a fast, physical game, with Lenny Wilkins running the point, dishing to a fierce front-court of Zelmo Beaty, Bob Pettit, Cliff Hagan, and Bill Bridges. Still, now that we have Jerry back, I feel we can outrun them and dispatch them quickly.

We win the first two games in L.A., then play as if we're all hungover and drop the next two in St. Louis. We blow them out in game five, 123–96 — I score 37 — then lose the next game back in St. Louis. The series tied, we come home for game seven. We win, 115–100, and take the series in front of a huge, screaming crowd, but these games have worn me out. My knees don't really bother me, but everything else in my body does.

And now we face Boston. Again.

We've taken the season series, 5–4, but the Celtics have the better overall record and earn home court advantage. We open with two games in Boston. We take an early lead in game one, but even though I put in 33, they have too much Russell. He outscores Krebs, Leroy, and Wiley, 25–5, and the Celtics pull off a 117–114 win. We lose game two, another frustration, 113–106, with Red firing up his victory cigar early in the fourth quarter, puffing away obnoxiously.

We come home to a completely sold-out Sports Arena. Local TV stations fight for broadcast rights, nobody wins, and a nearby arena shows the game on closed-circuit TV. As for the game, I don't care if they've got King Kong in the middle — I've had enough. Knee pain be damned, I score 38 points, yank down 23 rebounds, and hand out 18 assists, most of them to Jerry, who scores 42. We blast Boston out of the Sports Arena, 119–99.

The Celtics adjust. Russell again camps out beneath the basket, pummeling our three centers, who can do nothing with him. Heinsohn goes off for 35 and, although we hang in until the end, they nip us 108–105 to take a 3–1 series lead.

Do-or-die back in Boston. We do. I score 43, and we outscore the Celtics 126–119, to keep our hopes alive. Preparing for game six at the Sports Arena, I try to drown out all the noise that rolls at me: crowd noise, the noise of newspapers arguing over where basketball royalty resides — L.A. or Boston? — and the noise in the back of my head humming that we can't stop Russell.

The Celtics take the lead early. We come back, fall back, and come back again, but we can't quite catch them. Russell seems to be everywhere: a comet, flying in for tip-ins when a moment ago he was all the way back at half-court, appearing out of nowhere to block shots, hitting a key baby hook when he's seemingly blocked out. He puts on a basketball clinic, scoring 12 points, snaring 28 rebounds, and blocking nine shots, although that total seems low to me. I feel like he's blocked twenty shots at least, as he did while winning the NCAA championship years ago.

With time running out, Cousy picks up his dribble, clutches the ball to his chest, and, when the horn sounds to end the game—a 112–109 Boston victory, for their fifth-straight NBA championship—he flings the ball toward the rafters and runs off the court in front of the dazed and silent crowd. He has played his last game, exiting a champion. I congratulate him and search for Bill. The Celtics will be returning to Boston immediately after the game, so we will grab dinner another time. I find him and we hug.

"Congratulations," I say quietly, meaning it—and, frankly, resenting it. Hell, hating it. Losing hurts. Losing to *them* kills.

"Thanks," he says, and whispers, "Catch you in Boston, the basketball capital of the world."

. . .

I've had another fine season, averaging 34.0 points a game, second best in the league, behind only Wilt, and pulling down 14.3 rebounds a game, good for fifth place. I again make First Team All-NBA, with Oscar, Jerry, Pettit, and Russell, who wins the MVP award.

Both the national and local media, especially the L.A. papers, continue to write unbelievably flattering stories about me. I skim some of them, ignore most of them, and send all of them to my mother, who shares them with the rest of the family. Of course I'm happy that people write about me—humbled, really—but as I say, I don't care much about statistics and I get embarrassed by all the attention. I don't mind talking to reporters when I'm in the mood. When I'm not—which isn't that often—I duck them.

For me it all comes down to playing basketball at the highest level I can, always working at it, trying to improve, trying something new, and ultimately winning the game. I like to win. I *want* to win. It's not a high or anything like that, not the feeling you get from a drug, the way some people talk about it. But I do feel a kind of *rush*. And, yes, I would like to win a championship—I long to win a championship—but as long as Bill Russell runs with the Celtics, we can't seem to find a way to close the deal.

Over the summer, I take it easy, trying to take the pressure off my knees. For the most part, I stay close to home. I do say an official goodbye to the wild man, the guy who always makes me laugh, my friend Hot Rod Hundley, who's called it a career after eight NBA seasons and two bad knees. I attend a few charity functions, one a golf tournament—a sport I still haven't tried. And for the second or third year in a row, I work at Great Western Savings in the loan department. I like interacting with people and learning the real estate business. I can see that Los Angeles may be on the verge of becoming a boomtown.

At the end of the summer, before training camp, I visit my family in D.C. and attend a civil rights march, a massive three-day rally on the National Mall, in front of the Lincoln Memorial. I stand with hundreds of thousands of people as Martin Luther King Jr. delivers his stirring "I Have a Dream" speech.

What I see around me both inspires and saddens me. People shouldn't have to march on D.C. or give speeches or hold protests for something as basic and essential as equal rights for all.

· · ·

October 1963.

The NBA once again realigns, with the Chicago Zephyrs moving to Baltimore to become the Bullets, and the Syracuse Nationals becoming the Philadelphia 76ers. Maurice Podoloff, a well-known friend to the owners, retires, and J. Walter Kennedy, a players' guy, we hope, takes his place as commissioner.

The players formed a union about ten years ago, but frankly, because we've been stonewalled by the owners, we haven't accomplished much. We don't really *want* much, just what's fair. We're asking for a decent pension plan, a less brutal travel schedule, a larger meal allowance, and, as my knees continue to ache and burn, full-time doctors on every team. Perhaps inspired by all the protests and activism we see, we feel that our time has come. Well, most of us do. We elect Tommy Heinsohn from the Celtics as our union rep, and he sets up a meeting with the owners. The owners cancel the meeting. Tommy reschedules. The owners cancel again. Right before the season tips off, Tommy calls me. "If they keep dicking us around," he says, "we may have to do something big to get what we want."

"You mean what's *coming to us,*" I say.

"That's exactly why I'm calling you. You're the voice of reason. Every player respects you."

"What are you thinking of doing?"

"Boycotting the All-Star Game."

"If all else fails," I say.

"If all else fails."

Welcome to the sixties, I think.

"Bold move," I say.

"Would you support it?" Before I can answer, Tommy adds, "The players will do what you do."

I picture Wilt. "Not all of them."

"Most of them. So, will you?"

I imagine the hundreds of thousands of people standing in front of the Lincoln Memorial, all of them wanting equal rights, wanting what's fair, hearing "I have a dream."

"Yes," I say.

• • •

I don't want anyone to know.

I want to keep the secret from my teammates, from Fred, from Hoggy, and from every opposing player.

I don't want to tell Ruby, either. Ruby, pregnant with our second child, has quit her job. I especially don't want to tell her.

I don't want to tell anyone about the pain.

But as the season starts, I don't have to say anything, because everyone can tell. They can see. They watch me grimace as I come down with a rebound. They watch me start to sprint downcourt, then pull up and jog the rest of the way. They see that I sometimes step gingerly and favor my left leg because of a sudden shooting pain traveling through my right knee like an electric shock.

That's the strange part. I can go an entire half without any pain at all, and then suddenly — *flitt* — I feel as if someone has just driven a nail into my kneecap.

I can't get lift. I can't stop and go. I can't whirl to the hoop as violently or as fluidly as I want. My rhythm feels jerky. My timing is off. I don't care about my numbers, but the numbers don't lie. The numbers tell the story.

I averaged 34 points a game last season, *38* the season before.

This season, through our first seventeen games, I score 30 or more points exactly twice. I have games of 14, 13, 10, and 8.

Eight points?

I want to scream — not just in agony, but out of frustration.

People start talking.

After a loss, Jerry mutters something to a sportswriter about how ownership or the coach should be taking care of me and not allowing me to play and chancing a worse injury.

Other players — Rudy being the loudest — echo what Jerry says. Even Sam Jones of the Celtics says I should be allowed to rest. He worries that I'm in danger of getting severely hurt.

Fred says he wouldn't risk hurting me, because I'm so valuable to the team. But he doesn't play me fewer minutes or offer me a day off.

I say nothing. I want to play — with the pain, despite the pain. I refuse to give in to it. I keep playing. I feel I owe it to Bob Short, to the team, to the fans. But even some of them shout from the stands that I should sit and rest my knees. *He can't move! Sit him down!*

I block out most of the shouts from the crowd. But for some reason I hear this one.

I see doctors. Plural. So many I lose count. They present a spectrum of diagnoses and opinions. I have calcium deposits. I have tendinitis. I have cartilage damage. I have a disease. I have aging, twenty-nine-year-old knees that have logged a million miles on hardwood.

In other words, I'm old.

The doctors shoot me up with cortisone. They shoot me up with benzylidene, a drug they give to racehorses. They prescribe pills. I try acupuncture. I try heat treatments, ice treatments, massage therapy, x-ray therapy, every kind of therapy, including some kind of voodoo therapy I think they've gotten from a witch doctor.

I wear an Ace bandage, a special surgical wrap designed just for me, and a tight leg bandage wound around my leg that looks like a sleeve.

Nothing helps.

I score 16, 20, 23, and 8.

Eight points?

Then, on November 22, everything gets worse.

. . .

President John Fitzgerald Kennedy, who has spoken out repeatedly for equal rights, a president I campaigned for, someone who actually has given me hope, has been assassinated. I feel physically ill. And then I witness something I previously thought impossible. Our country — every citizen, black and white — grieves together. Time stops. Businesses close. The league postpones our games. I hole up at our house and sit with Ruby, glued to our television, watching nonstop coverage of the murder, then Jack Ruby shooting Lee Harvey Oswald, and then the funeral.

The next day, I join the team for our flight to New York as we begin a seven-game road trip. I take JFK's assassination personally. Staring out the window at an endless floating mattress of fluffy white clouds, I feel staggered, and I feel loss.

We beat the Knicks 119–112. I score 25, and I don't experience any pain in my knees. That night, I don't feel much of anything.

The next night, we play the Celtics in Boston. The moment I step onto the hardwood, I feel a stabbing in my right knee. I play through it, babying myself, being extra careful not to put too much pressure on it. At one point, the basketball squirts loose out of a scrum of players, and Sam Jones and I race after it. Normally, I—Rabbit—would beat him easily, no contest. But with my knee throbbing, I hold back, slow down, and allow Sam to grab the ball and drive in for an easy layup. Running back, he looks at me with genuine concern. "You all right?"

"Fine. Never better."

"Well, if your knees are acting up, I understand, but"—he nods at the other Lakers on the floor—"what's wrong with *them?*"

That I can't explain.

The Celtics humiliate us, burying us by 36 points, 114–78. I score 13. In the locker room after the game, Jerry, red-faced, sagging in front of his locker and staring at the floor, says that we embarrassed ourselves. He mutters that he's sick about how much my knees seem to hamper me.

I can't argue with him. I also can't continue like this.

So I push myself harder, and I change my attitude. I will no longer play through my pain. I will *fight* through it. I will smother it.

Two nights later, we play Detroit, a weak team. I score 37 and we bomb them, 127–111. I score on an arsenal of shots—jumpers, floaters, drives, dunks, and knifing spin moves that I come up with on the spot. Beating the Pistons seems to ignite us. We win three of the next four, and in that five-game span I average 25 points a night.

Then, after a two-point loss at home to the Warriors in which I score 32, Bob Short tells me that after our next game, against the Baltimore

Bullets, which we will play in Seattle, he is flying me in his private plane
to the Mayo Clinic, in Minnesota.

"Why?"

"Why? Your knees, that's why. Nobody can figure out what's going
on. The Mayo Clinic will. Best doctors in the world."

"I really—"

"You're going."

"We have games—"

"I know. You might miss one or two."

"I haven't missed a game in years—"

"You're *going*."

I decide not to argue. I do try to figure out how to get out of the
trip. I've never contradicted Bob Short before, but I have two major
concerns about flying to the Mayo Clinic: First, that the doctors will
recommend surgery. Second, flying in a private plane.

We beat the Bullets in Seattle, 134–120, and I score 29. Near the end
of the game, I come down with a rebound and land on somebody's foot.
Pain shoots through my ankle, and within minutes my foot swells to
double its normal size. I limp off the court and into Bob Short's private
plane.

As for my knees, the Mayo Clinic doctors don't recommend surgery
—yet. They identify the problem, definitively, as calcium deposits.

They think.

They're pretty sure.

Hell, they don't know, either.

They subject my knees to two days of state-of-the-art x-ray therapy, different from anything I've had before. I shove my legs inside a
boxy, high-tech machine that zaps some kind of electrical current that
pulses through my knees. After the treatment, my knees feel warm and
kind of mushy. The doctors tell me that I should come back for regular treatments. I can't see myself taking games off to stick my legs inside this Space Age contraption right out of *The Twilight Zone* while
my teammates battle the Boston Celtics. As we leave the Mayo Clinic, I

ask the doctor if he can prescribe some pills strong enough to knock me out.

"For the pain?"

"For the flight home," I say.

• • •

January 14, 1964.

The twenty best players in the league, plus every team owner, convene in Boston for the NBA All-Star Game. I've done my best to ignore the pain in my knees, which varies, depending on the day, from manageable to excruciating. I've refused to surrender to it, and in the dozen games leading up to the All-Star Game, my play has picked up and we've won five of seven games.

The weather forecasters predict that a massive snowstorm will hit Boston on Tuesday, so Jerry and I fly out on Sunday. On the flight, we talk about what awaits us — not the usual All-Star Game. We anticipate a battle beyond the game. The owners continue to refuse our demands for a pension plan and improved working conditions, and they keep dodging Tommy Heinsohn whenever he asks for a meeting. ABC has recently purchased a package for seventeen nationally televised NBA games, beginning with the All-Star Game, and Tommy wants to take action.

"No better time for a boycott," Tommy says.

"As long as everyone goes along with it," I say.

Jerry and I support the boycott. The truth is, most of us aren't getting rich playing in the NBA. Every player I know works a second job over the summer. What we're asking for — what we're demanding — represents a first, small step. A necessary step.

"We're like slaves to the owners," Jerry says.

Well, not quite, I want to tell him. But we do deserve a pension plan.

Tommy greets us in the hotel lobby with a handshake and a petition for us to sign. "This says you agree to boycott the game unless the owners give us our pension. Sign here."

"We're boycotting our first network TV game," I say, grabbing the pen from Tommy and signing my name. "This is either brilliant or suicide."

"We'll see."

"You got enough signatures?"

"It's gonna be close."

Extremely close. Eleven players sign the petition and nine refuse. An hour before the game, nervous and divided, the All-Stars split off into our separate East and West team locker rooms. Later, I'll hear that in the East locker room, Tommy, along with Bill Russell and Oscar Robertson, argued strongly for the boycott and for our rights, and gradually swayed the doubters. In our locker room, Wilt sits alone, sulking, his body language belligerent, his opinion contrarian. Half an hour before game time, Walter Brown, the Celtics' owner, comes into the East locker room and asks Tommy to withdraw the boycott. He argues that if we don't play the game, ABC will walk away from its contract and we will never again get a television deal. Tommy shrugs and says, "Let 'em."

"You will kill basketball," Mr. Brown says.

"I doubt that," Tommy says.

Mr. Brown loses it, screams at Tommy, and storms out. At that point, both teams bar the locker room doors.

A few minutes later, Bob Short tries to barge into our locker room. A Boston cop stationed outside blocks his way. Bob, not a fan of being told what to do or where he can and cannot go, starts screaming through the door, "You tell Elgin Baylor that if he doesn't get his ass out here fast, I'm done with him!"

Having had a year of pain and frustration, I move to the door and shout, "Tell Bob Short to go fuck himself!"

I don't know if he hears me. At that moment, I don't care.

I do know that everyone in that locker room knows how passionately I feel. The mood shifts then from division and uncertainty to solidarity, even with Wilt still silent and sulking.

With game time only minutes away, Tommy bursts into our locker room. "The commissioner wants to talk to us," he says.

We gather in the East locker room, all twenty of us, as Commissioner Kennedy tells us that he's spoken to the owners and that they have agreed to provide a pension plan and address all of our other issues. He asks that we play tonight's game in good faith. By a show of hands, we agree, 18–2.

We've won.

ABC televises the game and we change the NBA forever.

• • •

We limp through the rest of the season, at one point losing thirteen of fifteen games, Jerry nursing an injury that sidelines him for eight games. Despite my sore knees, I miss only two games, but we finish third in the West, with a 42–38 record, and fall to the Hawks in the first round of the playoffs. I end the year averaging 25.4 points a game and 12 rebounds, my lowest numbers since my rookie season. Even so, I again make the All-NBA First Team, with Wilt, Oscar, Jerry, and Pettit.

I vow to do better. I know I have a lot of basketball left to play, and I want at least one more shot at the Celtics, who go on to win their sixth-straight NBA championship.

Then summer comes, and with it three landmark events.

First, Ruby gives birth to our daughter, Alison.

Second, President Lyndon Johnson pushes through and signs the momentous Civil Rights Act of 1964. At long last, we have achieved equality, at least on paper.

Third, as I promised a year ago, I visit renowned orthopedic surgeon Dr. Robert Kerlan about my knees.

• • •

I agree to a consultation. That's all. I agree to talk to the guy.

I still believe I can manage the pain, although I admit that at times, my knees hurt so much I want to take a game off. Which I never do.

I drive to Dr. Kerlan's office feeling more than a little apprehensive. What will he tell me that seven other doctors haven't?

Elgin, be open.

Everyone says that—my mother and Columbia, on the phone from D.C., and my concerned L.A. contingent of Ruby, Fred, and the teammates I confide in, Rudy, Dick, and Jerry.

Okay, I'll be open. Totally open.

I enter the lobby and march over to the receptionist. I clear my throat and announce myself softly.

"Dr. Kerlan's expecting you," the receptionist says, gravely. "Go right in."

She nods at an open area I can see through an archway. An extremely hunched-over man wearing a suit and walking with two crutches clops into view and smiles. The guy looks like a question mark. Panicked, I turn back to the receptionist. "That's not—?"

"Dr. Kerlan? Yes."

"Oh, I, no, see . . ." I whip around and stare at the hunched-over guy. He acknowledges me by lifting a crutch and aiming its rubber tip at me.

This guy can barely help himself. How the hell can he help me?

A lump rises into my throat. I bend over and urgently whisper to the receptionist. "I just realized . . . I'm in the wrong place. Sorry."

"Mr. Baylor—"

That's all I hear.

I'm gone.

• • •

I go back.

I go back because I don't want to explain to my mother, Columbia, Ruby, and all the others why I never saw Dr. Kerlan.

Besides, I tell myself, we live in a new time. A new age. We should be trying to banish prejudice of all kinds. I shouldn't be biased against Dr.

Kerlan because of his physical condition. I decide to give him a chance and hear what he says.

Not only do I find Kerlan to be one of the smartest men I've ever met, but we become lifelong friends.

He examines me meticulously, putting me through a battery of x-rays and ultrasounds. He prods me, pokes me, and then isolates me in a dimly lit room and attaches me to machines I've seen only in science fiction movies. Finally, I sit across from him in his office. He stands behind his desk. He begins with a shocking statement that contradicts every doctor I've seen.

"You don't have calcium deposits in your knees," he says.

"I don't?"

"No. You have calcium in your quadriceps. Not an uncommon condition among football players, especially running backs and wide receivers."

"The skill guys."

He nods.

"What should I do?"

"No surgery, that's for sure. I recommend a special therapy we've developed. It involves heat treatment and pressure massage."

"How often?"

"Every day."

And so, every day, one of Dr. Kerlan's technicians hooks me up to a device that supposedly shoots pressurized air into my legs, the same way you pump up a tire—or a basketball. He also assigns me to a physical therapist, who works with me doing light weight training at his office. I miss one session all summer: the day I take my son, Alan, to the zoo. By Labor Day I feel pumped. My legs feel stronger, and the knee pain seems to have gone.

We start the 1964–65 season by tweaking the roster. Jim Krebs, Mr. Doom and Gloom, retires, taking his Ouija board with him. (The Ouija board prophesied that Jim would die before age thirty. A year after his retirement, while he is helping a neighbor, a tree limb will fall on Jim and kill him—at *twenty-nine*.)

We bring in six-foot-ten journeyman center Darrall Imhoff. In the draft, we also grab UCLA's outstanding guard Walt Hazzard, a clever playmaker who can both pass and shoot. We head into the season hungry and filled with hope. I warn Bill Russell at one of our preseason dinners that the Celtics' championship run is about to end. He laughs, his high-pitched cackle drawing looks in the restaurant.

"I understand why you feel that way," he says solemnly. "It happens to people of a certain age. The mind starts to go."

I have to laugh with him.

People of a certain age.

Well, it's true. I have recently turned thirty.

"It's okay, Grandpa," he says.

"What are you talking about?" I say. "You're older than I am."

"By seven months," he says. He pauses. "I do think about it."

"What?"

"The end. When it's over."

"I don't," I say. "I can't. Too much unfinished business. Besides . . ."

I look off and swallow the rest of the sentence. I turn back and fix my eyes on him.

"We are what we do." I hold a beat. "We are basketball."

• • •

I feel no pain.

I play at breakneck speed, but I tweak my focus. I rebound—hard—and run the break, dishing to our fast team of scorers: Jerry, having a banner year, Rudy, Walt, Dick, and Leroy, who has to be the fastest center in the league. I become the original model of what people now call a "point forward."

I get my share of points, too. But I don't feel I have to score in order for us to win.

We start strong, winning a bunch of games in October and November, including a victory over Boston by a bucket, 114–112, in which I score 36. We keep a steady winning pace the first two months of the season and head into the All-Star Game in St. Louis carrying a 23–18 record.

For some reason, even though we all play fast and loose like we used to on the playgrounds, this year's All-Star Game feels like a playoff game. Bill, the East's center, and Wilt, representing the West, go at each other, holding nothing back, matching rebound for rebound, block for block, and slam for slam, while Oscar, the East's floor leader, attacks the rim like a freight train. But we stay right with them, Jerry and me taking over, trading hoop for hoop.

In the fourth quarter, Gus Johnson of the Baltimore Bullets gets hot and pulls us within a point. I can't explain how Baltimore has become part of the West, but tonight I don't care. Finally, we fall just short and lose, 124–123. I score 18 and contribute several assists on fast breaks. I never like losing to Bill — I never like losing, period — but I don't take the All-Star Game all that seriously. In the West locker room afterward, though, Wilt seems pissed and leaves without speaking or showering. A day or so later, the Warriors trade Wilt to the Philadelphia 76ers for a couple of bench players and some cash. They give him up for practically nothing. I'm shocked. But somehow I'm not surprised.

Through February and into March, we dominate the league, winning sixteen of eighteen games, including an inspired victory against Boston at the Garden. Down by 13 at the half, we storm back and stun the Celtics 104–102. I play a strong all-around game, sweeping the boards and scoring 22.

As the season winds down, we've become Hollywood's newest heroes, selling out the Sports Arena, with prominent celebrities now season ticketholders. When I go out at night, I hobnob with Hollywood royalty. The truth is, to my surprise, famous people actually seek me out. I meet Joe Louis and Muhammad Ali. I become friends with Willie Shoemaker and Maury Wills. A few times I even go to a nightclub with Frank Sinatra, who can be charming and funny until he has a few drinks. Then he gets loud and aggressive and starts looking for fights.

A couple of weeks before the playoffs, the grind of the regular season starting to take its toll, my knee pain returns. At first I feel a subtle but persistent throbbing, then a constant ache that burns through my knee-cap, then a sharp, shooting pain that causes me to sit out a game. Then, against the Pistons, I bang knees with the guy guarding me. I somehow finish the game, but sit out the next two. We finish the 1964–65 regular season with a record of 49–31, first in the West, and I put up strong numbers, averaging 27 points and 13 rebounds a game. Painful knees and all, I again make the All-NBA First Team, this time with Oscar, Jerry, Bill, and Cincinnati's Jerry Lucas.

• • •

April 3, 1965. The Los Angeles Sports Arena.

Game one of the 1965 playoffs, against the Baltimore Bullets.

More than sixteen thousand fans pack the stands. The arena thrums with excitement. Slick Leonard, my former teammate, coaches the young, rough, athletic Bullets, led by Walt Bellamy, Johnny Green, Gus Johnson, and Terry Dischinger. I badly want a rematch with the Celtics, but I would never look past this bunch.

We shake hands for the center jump, Bellamy tips the ball to Dischinger, and we retreat to defend our basket. I've already made up my mind that I want to start fast. I'm determined to score quickly. I want to knock Gus back on his heels before he starts putting his hands all over me and mauling me as I drive to the basket.

The Bullets miss their first shot, Leroy snags the rebound and flips the ball to Jerry, and the rest of us bolt downcourt. I set up on the right side of the lane and receive a bounce pass from Jerry. I peek over my shoulder, locate Gus, who settles in behind me, fake once, twice, start my dribble, spin, step, and —

My left kneecap explodes.

A wildfire roars up my leg. I crash onto the floor. The world goes dark.

I hear something that sounds like shock.

I hear that: the *sound* of shock.

I blink, and the darkness shimmers into filmy focus, and then rough hands grip me under my arms and hoist me off the floor. I gasp. The pain that rips through my leg sends tears gushing down my cheeks. I sniff and I step. Once. Only then do I make out Hoggy's round form as he helps me walk, guides me — *one step, one step* — my left leg rigid as a wooden plank, but airless, without form, as if it no longer exists.

"Hoggy —"

"It's okay, Elg. Lean on me."

One step, one step.

In the distance, I see fear on people's faces. I see entire rows of people standing, their mouths open, as if they are the background in some bizarre cartoon. I feel my tears bubbling and then dripping into my mouth.

One step, one step.

"I got you. Easy. Take it slow."

Silence. The world has gone mute. Someone — God? — has turned off the sound. I raise my head, narrow my eyes, and lean back, the tears puddling, the pain raw, hot, as if someone has sliced my kneecap with a razor. I approach the sideline — *one step, one step* — and I think, ridiculously, *I'll walk it off. I'll be okay in a few minutes.*

Hoggy eases me to a spot in front of my teammates, who are clustered around us, and deposits me onto the bench. I look at him and see a grimace on *his* face and fear clouding his eyes. I inhale deeply and blow out the air as slowly as I can, trying to release the pain with my breath.

I look at Hoggy. "I'll be all right. Give me a few minutes." I nod to reassure us both. "I fell awkwardly. That's all. I'm gonna test it."

I stand, and my leg shakes uncontrollably. I thrust my arms out. Hoggy catches my hands and steadies me. I breathe again, silently count to three, and take a step. My leg convulses. I stop, look at the floor, and then decide to jog along the sidelines. I will run this off. *Run* it off.

In a second. Not yet.

"You ready to go back in?"

Fred. He looks at me and I see fear dancing in his eyes, too.

"You just tell me when," he says, his voice slightly above a squeak.

I nod. I take a step.

One step . . .

And then I run.

I run along the sideline, pull up, and then I run back. I stop and gasp. I hold my breath to keep myself my crying.

"I . . . I . . . *can't*," I say to Hoggy, to Fred, to myself.

"Let's go into the locker room," Hoggy says, grabbing me under the arms again.

I inch away. "Don't . . . help me. Let me . . . walk on my own."

I take a step—*one step, one step*—and slowly, on my own, I walk toward the locker room.

The crowd erupts. I acknowledge the roar with a wave.

I think.

I think I make it into the locker room on my own. I think I wave at the crowd. Or maybe I dream it all. I can't be sure. I barely know where I am. The pain floods me, robs my senses, erases my thoughts.

"Fuck!" I scream, my voice echoing through the corridor. I don't know if I say that aloud or in my head. After that, I remember only random images: An ambulance. My hands gripping a metal bar, my body flailing, the pain sawing into my kneecap. Dr. Kerlan leaning over, pressing papers into my hand for me to sign. My fingers trembling, the pen in my hand shaking. A glass held to my lips. Swallowing pills. Someone in white pushing a needle into my arm.

Dr. Kerlan again. "I just read the x-rays."

X-rays? When did I have x-rays?

"Your left patella—"

"English," I say.

"You've torn off the top half of your left kneecap."

I see white sheets, white walls, white clouds. I lose track of time. I lose days.

I wake and my left knee itches. I try to reach for it. I can't. I try to lift my leg toward me. I can't. I can't raise my left leg at all. It feels as if my leg is made of cement. Then I realize that my leg is encased in a cast from my ankle to my hip. The cast is suspended above the bed in a sort of hammock attached to a pulley. Dr. Kerlan stands next to the bed, wearing a white smock, a surgical mask dangling around his neck. He looks me over. "How you doing?"

"Next question."

"I have a present for you, a souvenir." He hands me something that looks like a jagged vanilla wafer the size of a quarter.

"What's this?"

"A calcium chip. I found it floating around inside your knee. At least that won't be bothering you anymore."

I stare at it.

Dr. Kerlan presses my shoulder gently and says, "So, I excised the upper portion of the patella—"

"English."

"I cut your left knee open, removed the torn patella, repaired it, and put it all back together."

"I . . . I . . . need to know . . ."

I can't finish the sentence. I can't speak. For several moments, I feel as if I can't breathe. I blink at Kerlan, my eyes filling up. "Will I—"

"I don't know if—" Dr. Kerlan says softly, then clears his throat and starts again. "Elgin, I don't think you'll ever play again." Then he says quickly, "If you are able to play, it'll be a miracle, and you won't be the same player. You won't be able to do the same things. I don't know how well your left leg will heal, how strong it will be—"

My stare cuts him off. "I was asking if I would be able to walk again."

He studies the floor and then meets my eyes. "Yes. You'll be able to walk."

"Will I be—normal?"

I look at Dr. Kerlan, bending over my hospital bed, his hunched back suddenly seeming so much more pronounced, and I feel my face flush. I shouldn't have asked him that.

"I'm sorry," I say. "I didn't mean—"

"It's all right," he says. "Yes, I think you'll be normal. You may have a limp."

"A limp."

"Possibly."

A limp.

Can I play basketball with a *limp?*

What has happened to me? Am I done? Have I come to the end? *Is my career over?*

I flash forward to . . .

To what?

I'm thirty years old. An NBA All-Star. Multiple times. Every year. All-NBA First Team. Multiple times. Every year. And now I will do . . . *what?*

Sell suits at Nordstrom?

Greet customers at a Las Vegas casino?

Move back to D.C. and teach high school gym at Spingarn?

I force myself to sit up in bed. Pain slices through my leg. I wince.

"No," I say.

It can't end like this. It's too soon. I have to keep playing. It's all I know. Basketball is who I am.

I will rehab. I'll do whatever it takes. I'll come back. I may not be the same player—I'll live with that. I'll be a different player, a new player. I'll create a new normal. I will fight back.

"I'm gonna play again," I say to Kerlan.

"Well, if anyone can do it, you can," Dr. Kerlan says. "I mean that."

"I'm not done," I say.

I will come back.

I am not done.

9

HOLLYWOOD (ACT II)

IT'S NOT THE GETTING AROUND THAT BOTHERS ME.

I move like a champ, the king of crutches. The kids, Ruby — everybody is in disbelief at how fast I motor along on my sticks. I wish I could set up a leg-in-a-cast race against other athletes. I'd win in a walk. Make a fortune betting on myself. But small problem:

I can't get dressed.

I can put on shorts. I pull them over my cast and wriggle into them with only minor frustration and occasional cursing.

But I can't get my damn *pants* on.

It's a comical nightmare. I refuse to allow Ruby to help me. I have too much pride. I insist on getting dressed myself. I close the door to the bedroom.

I try various methods.

I sit on a chair. Terrible idea. You cannot pull on a pair of pants while sitting on a chair. You get no traction, and unless you're a contortionist, you can't reach over far enough to force your cast into a pair of pants. I try for fifteen minutes, curse, fling the pants against the wall, and tip over the chair. I lean against the bed — worse than the chair.

I stand. I fall over.

I use the tipped-over chair to pull myself to my feet.

I lean against the wall. Success.

Eventually. Takes forever. And I am a man of little patience. I force myself to go slowly. I pull my pants up inch by inch and then the fabric bunches up around my calf and sticks there, a woolly glob.

Then I try yanking my pants up fast. I fall over.

I consider ordering special pants with a slit up the side.

I calm myself. I murmur aloud: "Six weeks. That's all. Six weeks. Eight weeks max."

And then, alone in the house, I scream.

· · ·

Time blurs.

Truthfully, I've tried to block it all out. I don't want to remember. I want to leapfrog in time from a certain point in the past, when I ran and played with abandon, when I didn't consider myself a one-legged, half-crippled basketball player, to a moment in the future when I return to being myself, flying up and down the hardwood at full speed, blasting off at the free throw line and leaping, going airborne. I want to erase the *now*. That's what I want. I want to return to *myself*.

I know it's too much to ask.

And so I live in a perpetually unsettled state of anxiety, semidepression, extreme anger, frustration, and fear. I ask myself too often:

Will I play again? Will I play well enough to compete at the same level? Will I play well enough to satisfy my standards?

I feel so impatient. I want to start rehab—and I want rehab to be over. Time just crawls. I clop around the kitchen, watching the summer flit by.

I keep up with the news. I read the papers and watch TV. Players check in with me. I hear rumors that Bob Short wants to sell the team. The rumors start sounding like more than rumors.

We draft Gail Goodrich, a six-foot-one guard from UCLA, Walt Hazzard's college running mate, who's tough, quick, a good outside shooter.

We trade Dick Barnett, an elite scorer, the guy Chick Hearn calls

"Fall Back Baby" because of his unique shooting style, to the New York Knicks for Bob Boozer, a six-foot-eight banger, the former All-American from Kansas State. I remember Boozer as the guy who inadvertently elbowed me and cracked my rib during that NCAA Final Four game.

I'm not sure how I feel about the trade. I like Dick personally, enjoy hanging out with him, love how he dresses, and I hate giving up the scoring. And then I wonder—

Did the Lakers trade for Boozer as an insurance policy in case I can't play?

Oh, man, I can't wait to get to *work*.

●   ●   ●

Ruby and I buy a big house in Beverly Hills, high above Sunset Boulevard. On a clear day, you can see all the way to Catalina Island from our living room. I buy a new car, a Jaguar, and we adopt a couple of German shepherds we name Brutus and Caesar. The house features a big backyard where the dogs and kids can romp and a huge country-style kitchen with a center island.

The house that basketball bought is about to become the house where I will rehab.

Six weeks after the injury, Dr. Kerlan cuts off the cast. I stare at my left leg. It looks emaciated, about half the size of my right leg. It looks like a chicken leg.

"You have to take it slow with rehab, Elgin," Kerlan says. "Pace yourself."

"I know. I will. I just have to be ready for training camp."

"Seriously," Kerlan says. "Don't push yourself."

"You know me."

I push myself.

I begin every morning in the kitchen. I sit on the kitchen counter, lash a forty-pound weight to my knee, and do sets of leg lifts. I jump rope in the backyard. I ride a stationary bike. I run laps, building up to three miles a day.

Slowly, my leg returns to normal size. My leg feels foreign, wooden, and I can't bend it completely, but I see no trace of a limp. I wonder if I will be able to stop and start and change direction the way I used to. If I can't, what will I do instead? I'll have to compensate. Use trickery, guile, rely on head and shoulder fakes. And I worry about jumping. I never thought about jumping before, but I don't see how I will ever get the same spring. I've always used my left leg to launch myself. I no longer experience knee pain, but my leg doesn't feel nearly as strong.

Before training camp, the rumors come true. Bob Short sells the Lakers, for more than $5 million—an outrageous price—to Canadian entrepreneur Jack Kent Cooke. New rumors start, the juiciest being that Mr. Cooke wants to break ground immediately on a brand-new arena. Cooke, an owner of the Washington Redskins NFL team and now a resident of Los Angeles, *loves* hockey and plans to bring an NHL team to the city. Hockey in Los Angeles? I don't see it.

I continue my rehab through the summer. Each day, my leg feels slightly stronger. Then, one day in mid-August, I stare at the TV as Watts, an almost entirely African American section of the city, not far from the Sports Arena, bursts into flame. Over the next six days, I stay riveted to the coverage of the Watts riots, witnessing unspeakable violence, protests, looting, fires, explosions, police and National Guard troops in riot gear swarming the streets of South Los Angeles, tanks rumbling through swaths of protesters. I watch all this feeling stunned and heartsick as police and law enforcement kill thirty people and arrest thousands. It seems as if I'm watching war footage. *What country do I live in?* I think as I watch blocks of buildings burning to the ground. *What can I do?*

I scan my spacious country kitchen and take in our state-of-the-art appliances, stereo system, television set. I peer out the window at our wide patch of lawn, green as a golf course, spray from our sprinklers arcing languidly, dusting the grass like stardust. I picture my neighbors, all of them white, secluded, safe, so far removed from the riots in Watts, and from the daily reality of poverty, pain, and hopelessness.

I don't know what to do. *What can I do?* I'm just a basketball player.

• • •

Training camp starts, and I'm not right. I try to run like before, but my left leg is stiff as a plank. Half the time, I feel as if I'm dragging my leg behind me. I also feel clumsy. I've never felt clumsy. Worst of all, I feel tentative. I'm afraid to go all out. I don't want to reinjure myself.

It's training camp, I think. *I shouldn't push myself.*

Fred feels the same way. He excuses me from the most strenuous drills and allows me to scrimmage for only a few minutes at a time. Meanwhile, before practice and on my off days, I stretch and limber up longer than I ever have. I still feel stiff. Finally, Dr. Kerlan gives me permission to play in an exhibition game.

I play at half speed, and I stink. I baby myself. I put my left foot down lightly, as if the court is made of thin ice. I am able to find my jump shot, but I don't really drive and I can't play defense at all. Guys blow by me.

Is my leg that weak? Or is it all in my mind? Or both?

It doesn't matter. It's training camp. Right?

• • •

Dr. Kerlan clears me to play in the season. He tells me that I'm a miracle. He admits that he underestimated me. He swears he will never make that mistake again, especially since he and I have formed a weekly poker game, along with Willie Shoemaker, Walter Matthau, and a few other guys.

We begin the season on a four-game road trip that starts in San Francisco against a young, raw, but talented Warriors team led by All-Star guard Guy Rodgers, big, mobile center Nate Thurmond, and the second pick in the 1965 draft, Rick Barry, a six-foot-six forward who can score, rebound, and pass. Some people compare him to a young me.

I don't like that. I don't like comparisons. I'm still me. I'm still here.

But in this game, I worry about moving too quickly or too abruptly. Rick, who later tells me he has modeled himself on me, seems nervous in his first NBA game. We beat the Warriors 122–115. Rick scores 17, while I score a grand total of eight.

We beat the Hawks the next night. I play it even safer and stay outside. My jump shot works, and I score 22. We continue to Boston. Red Auerbach has begun the season with surprising news: this season will be his last. He announces his retirement with a challenge that he directs at us, saying that this will be our final shot at the king. He hints that beating Boston without him at the helm won't carry the same weight. I see this as the usual Red psych job. He looks for any advantage. This night, we play them tough but lose by four. I put in 22 points, again roaming mostly outside and trying to keep pressure off my left leg.

On to New York. Warming up before the game, I feel . . . off. Unsteady. My knee doesn't exactly hurt, but it doesn't feel right, either. Fred limits my time, and I score exactly five points in a 106–101 loss.

In our home opener in Los Angeles, in front of a huge crowd and our new owner, I push myself. I score 18 and we beat the Knicks by two. We play the Knicks again the following night and we pour it on, blasting them 140–112. Fred rests me for much of the game and I score only 12.

For now, the team dynamic has changed. With Jerry shooting the lights out, Walt Hazzard and Gail Goodrich running the team, and Rudy banging the boards and scoring on short-range jumpers, I find I can play a secondary role while not putting too much pressure on my leg. Because that's what I do. I hold myself back. I don't do it consciously, but I'm aware that I'm using my right leg more.

I realize I'm overcompensating, because pain starts shooting into my *right* knee. *Calcium deposits,* I figure. Often when we play, I feel as if I'm playing on one and a half legs instead of two. I deal with it. Through the first eighteen games, I settle into this supporting part, allowing the others to do the heavy lifting. It works. We get off to a strong 12–6 start.

In late November, once again playing San Francisco, I do something stupid. I pivot away from my defender, stop abruptly and start suddenly, and feel a pop in my right knee. Here it comes. The *pain.* Roaring into my knee. I collapse onto the floor, both hands cradling my knee, frustration and anger rising into my cheeks.

"Sprained ligaments," Dr. Kerlan says the next day. "You put so much stress on your right leg that it gave out. Luckily, you didn't tear anything."

"I don't want to ask," I say.

"A month."

"A *month?*"

"You still have your crutches, right?"

"Yes."

"You'll need them."

He then oversees his assistant, who constructs an ankle-to-hip cast over my right leg.

Fred places me on injured reserve and I watch the next twenty games in street clothes, sitting behind the team.

Cast off, I return around Christmas, wobbly, uncertain, and out of shape. For the first time in my life, I come off the bench. I seriously start to consider whether, at age thirty-one, I have reached the end of the line.

On January 2, 1966, we play the Celtics at the Sports Arena. We lose 124–113. I play a few minutes in each half and score two points. Bob Boozer, starting in place of me, scores 17.

• • •

January 11, 1966. Cincinnati, Ohio. The annual NBA All-Star Game.

For the first time in my career, I have not been invited.

In the midst of a five-game road trip, Fred, once again the coach of the West All-Stars, and Jerry and Rudy, this year's Lakers All-Stars, fly to Cincinnati after our game in Detroit. I fly with the rest of the team to Boston, where we will play our next game. Even though we have two days off, I don't feel like doing anything. I crash in my hotel room, order room service, and mindlessly watch a nature show on TV. Something about a wounded lion. Perfect. The phone rings. I grab it.

"How you doing?"

Dick Barnett. My friend and former teammate, now a New York

Knick. Dick has found a home in New York. For the first half of the season he has lit up the league, averaging 24 points a game, third best in the NBA. The league, though, has snubbed Dick and not selected him to the All-Star Game. We both know why. In some sense, I have been expecting this call.

"I'm okay," I lie.

"How's the leg?"

"Which one?"

"I hear you," he says. Dick knows my frustration, my uncertainty. "You'll be all right. You'll figure it out."

"How you doing?" I say.

"Well, you know."

I do know.

"You deserve to be in Cincinnati," I say quietly.

"Thanks," he says. "Not sure what more I could have done."

"You did enough," I say, surprised to hear the heat rising into my voice. "You did more than enough. Only two guys in the league are averaging more points than you: Wilt and Jerry."

"I guess . . ." Dick stops. "We don't have to talk about it."

We do, I think. *We should. We all should.* So I say what we both feel, what we both know.

"The league didn't want to invite another black player to Cincinnati. Got too many black players in the All-Star Game already. Not enough white faces for the crowd, for TV."

Silence. I hear Dick breathing, but neither of us speaks.

"We both know that's the way it is," I say. "I know they picked at least two players over you just because they're white."

"What do I do?"

"The only thing we can do."

I pause.

"Be better," I say.

"I thought third-best scorer in the league was pretty damn good."

"It is," I say. "If you're white."

I hang up with Dick, turn off the TV, and stare out the window. The phone rings again, jarring me. I answer it and hear Fred Schaus's voice. He sounds strangely upbeat. "I just spoke to Kennedy," he says. "You're coming to Cincinnati."

"What?"

"Lou Mohs and I worked it out. We arranged for you to sit on the bench. You didn't make the team because of your injuries and the month you missed, but the NBA wants to honor you anyway. It's a tribute to you, to what you mean to the league." I can picture him beaming through the phone. "You deserve it, Elgin."

Dick Barnett deserves it, too.

Images then rush at me. I see myself years ago in that locker room in Charleston, West Virginia, Hot Rod Hundley standing next to me as I tell him I won't play in his hometown.

Baby . . .

I see Watts on fire, tanks plowing through rows of protesters, police in riot gear, their guns raised, charging like storm troopers.

Don't . . .

I see my father's strap, Columbia's tears. I hear her screams.

Play . . .

I hear Hot Rod's voice.

"Baby, don't play."

I'm only a basketball player. What can I do? Not much. Except what I can.

"Thank you, Fred," I say into the phone. "And thank Lou and the commissioner. But I'm going to stay here in Boston."

"You're not coming to the All-Star Game?"

"No."

"Are you sure?"

"Positive," I say. "I'm gonna sit this one out."

• • •

At the end of January, back home, we take on the Baltimore Bullets, playing two games in three days. On Friday night, we beat the Bullets

138–123, and we follow that up by beating them Sunday afternoon by an almost identical score, 136–122. I monitor myself in both games, being very cautious about how I cut, where I drive, when I jump. At one point in the second game, Gus Johnson, the Bullets' All-Star forward, steps back and slowly shakes his head. "He's got no legs!" he shouts to his teammates.

Normally—before my leg injuries—if Gus or anybody else said something like that, I'd torch him. But I say nothing—and I do nothing. He's right: I've got no legs. At least I don't have the legs I used to.

I mope to a total of nine points.

After the game, I get a message that Dr. Kerlan needs to see me in his office first thing in the morning.

What now? Did some x-rays come back with another awful result?

I arrive late to Dr. Kerlan's office, almost unheard of for me. I'm never late. But if Bob is about to deliver bad news, I'm in no hurry. I take a seat across from him. He folds his hands on his desk and looks at me for a long time. "I was wondering when you were going to tell me," he says.

I shift in my chair. "Tell you what?"

"That you retired." He leans across the desk. "If you're not going to play full out, give it everything you got, you might as well retire. Take a seat next to me. Become a season-ticket holder. I can get you seats next to Doris Day."

I lower my eyes and stare at his fingers, linked together.

"Let me show you something, Elgin."

Bob opens his desk drawer, takes out a copy of the *Los Angeles Times,* and starts to read. "'Elgin Baylor is poetry in motion. Why would anyone go to the Royal Ballet when they can see him?'" He slaps the paper on his desk. "You know who said that? Jack Kent Cooke. He bought the Lakers because of *you.*"

"I'm not the same player now."

"No. Probably not. But you might be close. You'll never know unless you test that knee all the way."

I lean into him. "How do you know my knee won't give out?"

"It won't."

"How do you *know?*"

"Because *I* did the surgery." He allows a moment for that to sink in. "I know what I did. The knee's strong."

"I don't think I can do what I used to."

"Don't think. *Do.*" He pauses. "I never thought you'd play again. I told you that. Now I'm saying you can not only play — you can play just as well."

"I know I can't."

"Fine. *Almost* as well." He unfolds his hands, taps his fingers on the desktop. "Name one player in the league who can guard you. Right now. Today. With your knee. Time's up. I'll tell you who: nobody."

Bob sits back in his chair and says, quietly, "Elgin, your knee's not holding you back. Your head is."

· · ·

Everything shifts.

Could be because of Bob's pep talk, or because I finally feel fully rested and in true game shape, or because I start to figure out a different way to play. Could be all three; I don't know. But in the last ten games of the season, I average 27 points and 14 rebounds a game. Before we tip off against Baltimore, I grin at Gus Johnson and tell him that, unfortunately for him, I now have two good legs. I destroy him and the Bullets for 37 points as we beat them 126–105. I pour in 46 against the Knicks, though they nip us in overtime, and 30 in a win against Detroit. We finish with a 45–35 record, again first in the West. Despite my late-season scoring outburst, I average only 16.6 points a game, a career low.

I'm rarely completely pain-free, but I learn to ignore any sudden pangs I feel. A twinge here or a throbbing there no longer affects me. Before my conversation with Kerlan, I interpreted the slightest knee or leg discomfort as a warning signaling certain danger and I would slow down, or rely on my other leg, or play at half speed. Now I go all out. I don't pace myself. I keep my foot to the floor. Same as I always did.

But in one way, I do the exact opposite of what Kerlan said.

I don't think less. I think more.

I break down every defender, situation, tendency—every moment I'm on the court—and I calculate how best to exploit every opportunity. I used to be jazz—improvising wildly, feeling my way through each game without thought or worry. I just *played*. Now I approach each game with a set list: nothing too specific, but at least a semi-plan. If I know I'm stronger than the guy guarding me, I won't get cute and try to throw him off with pump fakes and a fancy flip shot. I'll just overpower him. If I know I'm quicker than the guy guarding me, I'll just blow by him and take a layup. I don't need any extra sauce. And if two guys double-team me, I won't try to dazzle them with a move I make up on the spot. I'll find the open man. By necessity, my injuries have redefined the way I play. I now take what teams give me instead of taking whatever I want.

. . .

April 1, 1966. Nearly a year to the day since I tore my kneecap in half.

We take on the St. Louis Hawks at home in game one of a best-of-seven semifinal series. I score 22 points and we blow out the Hawks, 129–106.

With no sign of pain in either knee, I hit for 42 in game two, in front of a grinning and applauding Bob Kerlan, and we win again, 125–116. Up 2–0, we head to St. Louis. I feel confident and loose and relaxed for the first time in a year.

Too confident, too loose, too relaxed.

We drop three of the next four games, forcing game seven in L.A.

Game seven evolves into a typical Hawks–Lakers slugfest. We batter each other for three quarters and head into the last period dead even and dead tired. Then, somehow, we pull away, finally beating the Hawks 130–121 and taking the West. I score 33.

Now, to Boston. Here we go again.

In front of a wild Boston Garden crowd, we get off to a hideous start in the first quarter of game one, trailing the Celtics by 14. Then we shake loose and dominate the second period, closing to within one at halftime. We trade baskets the entire second half, catch them, and head into overtime. We take them out in the extra period. Satch Sanders and John Havlicek take turns guarding me. Feeling strong, feeling *renewed,* I dazzle with an assortment of new moves, swirl by them, dunk over them, zip along the baseline and dip under them. Havlicek fouls out, I go for 36, and we shock the Celtics, 133–129. After the game, Red Auerbach, who announced his retirement at the beginning of the season, hits the basketball world with yet another stunner.

He will be replaced as coach by Bill Russell.

Bill will be a player-coach, and the first black head coach in NBA history.

Leave it to Red to steal the spotlight from us.

I'm not sure when Bill found out—he tells me at dinner that he never expected Red to make the announcement during the playoffs—but for the next three games, Russ plays out of his mind. Of course, we have no one to match up with Bill, and the Celtics crush us in each of the next three games, taking a 3–1 series lead. No team has ever come back from a 3–1 hole to win an NBA Finals. And now we have to play the Celtics in Boston. I promise myself that they will not win. They will have to beat us.

I score 41 points, and we beat them in Boston 121–117.

As we return to L.A. for game six, trailing 3–2, the press goes a little crazy. Somebody interviews Schaus, who tells the reporter that, given the choice of any player in the NBA, he'd take me over anybody. Dr. Kerlan gushes to another reporter that I've made an amazing recovery, admitting that he never expected me to come back the way I have.

It's all fuel.

Jerry, Rudy, Gail Goodrich, and I each score more than 20 points and we win game six, 123–115, to force a game seven in Boston.

Of course, by now you know what happens.

We lose.

I go cold, Sam Jones gets hot, Bill Russell turns into some kind of human kraken, his arms whirling tentacles, as he scores 25 points, devours every rebound in Boston Garden, and blocks any shot he wants to. The scoreboard reads 95–93 Boston, but the game is never really close. The game ends and a river of fans floods the parquet floor to celebrate their eighth championship in a row. The Celtics have won another heavyweight fight, and once again I feel like their helpless sparring partner. Or, worse, their punching bag. As Red fires up his final vile victory cigar, I feel thankful for one thing.

I'll never have to watch him do that again.

• • •

October 1966.

I can't wait for the season to start.

At the beginning of the summer, Bob did some touch-up work on my knee, drilling pinholes, threading and securing what's left of the patella, reattaching and refitting some tendons from another part of my leg to my kneecap. Tightening everything up, he said. Bottom line: I feel stronger than I did last year. More than anything, I want to put last season behind me. I want to move on.

The NBA, too, seems to be moving forward. We've added a new team, the Chicago Bulls, which cost us Bob Boozer in the expansion draft. The players' union, led by John Havlicek and Oscar Robertson, also keeps working toward improvements, better pay, and increased benefits. Among the items on the agenda: flying first class instead of coach. I don't care about the upgrade you get in service or in food — I care about the upgrade in legroom. It's ridiculous for basketball players to fly across the country stuffed in coach like sardines, our knees jammed up into our chins for seven hours.

And the league moves the Baltimore Bullets from the Western Division to the East. That always bothered me. I know it's no big deal in the scope of things, but it shows me that at least the league is starting to think straight.

Early in the summer, Lou Mohs, our general manager, and Fred make a bunch of changes to our roster. Having been pushed around by stouter teams such as the Hawks and the Celtics, we bulk up. We trade sweet Leroy Ellis, who's fast but slight, to the Bullets for Jim "Bad News" Barnes, a bruising six-foot-eight forward who I hope lives up to his name. We bring back Tommy Hawkins, a small forward who can still jump and rebound. Then, in addition to our first draft pick, hot shooting guard Archie Clark, out of the University of Minnesota, we draft seven-foot center Hank Finkel, from Dayton University, and six-nine John Block, from the University of Southern California. Then we trade for Mel Counts, the former Celtic, another seven-footer.

I get what Lou and Fred have in mind. Go big and go big in numbers. Between Mel, Hank, and our returning six-ten starter Darrall Imhoff, we will throw three giants at you. Size matters in a league that features Wilt, Bill, Nate Thurmond, Walt Bellamy, and Willis Reed. We have no one center who can match up inside with any of those guys, although all three of our centers can shoot from fifteen feet out. That should pose a problem for opposing centers, or at least force them to make a decision: give this guy the outside shot or vacate the block and go outside to guard him, leaving the middle open?

I start training camp feeling like a new man. Actually, I feel like the old me. Except that now I know I have certain limitations—I don't have as much launch when I jump, I don't cut as crisply, and I never forget I'm thirty-two.

•   •   •

We open the season on October 15 without Jerry, who has suffered another injury that will sideline him for at least seven games. Every team has injuries—part of the game. We just seem to have more than most. *Maybe I can carry us until Jerry gets back,* I think. I play like a force against Baltimore, spoiling their home opener, beating the Bullets 126–115. I pump in 36, spinning, stopping on a dime, and drilling jumpers, or whirling and driving to the basket, leaping, and flipping up a no-look layup.

See that? I want to shout to my bewildered and flat-footed defender.

One writer would call me "The Godfather of Hang Time," which makes me laugh.

I can't hang in the air longer than anyone else. It's not physically possible. Gravity prevents such a myth. Here's the secret to my ability to appear airborne longer than other humans: a shrug, a shoulder wiggle —some call it a tic—a fake, and a super-quick first step. Then I get the defender to jump *first*. That's all.

Hang time. You might as well believe in Flubber.

I jog into the locker room after the Baltimore game, feeling we had it going on tonight, like I got dealt four aces to your full house. We stole one on the road. I clap my hands. *Let's keep it going.*

Three nights later, on October 18, we arrive in New York for the Knicks' home opener in Madison Square Garden, a game we lose, even though I score 31. I remember this game mostly because of "the Brawl."

During the third quarter, I get clubbed on my way to the basket. The ref blows his whistle, a rarity on this night, and I walk over to the foul line to shoot two free throws. The game up to this point has felt like a back-alley fight between Willis Reed and Rudy LaRusso, who have exchanged words and elbows since the tip, both of them constantly stalking and complaining to the referees, Mendy Rudolph and Richie Powers. Mendy and Richie want no part of their beef. That's obvious, because they keep turning their backs and walking away without talking to them or teeing them up. At one point, I hear one of the refs say to either Rudy or Willis or both, "Just shut up and play."

I sink my first foul shot, then dribble once, bend at the waist and arc in my second shot, vaguely aware of Rudy and Willis jostling and shoving each other. I see bodies hurtling and colliding, elbows flying, players running upcourt, then Rudy trips and sprawls onto the floor, springs to his feet, and takes a swing at Willis. Willis, massively muscled, a southpaw, clearly has had some training. He ducks Rudy's punch, unleashes a looping left hook, and punches Rudy in the face.

Seeing this, Darrall Imhoff runs up behind Willis and throws his arms around him to prevent him from hitting anybody else. Willis busts free and belts Darrall in the jaw. Then Willis goes berserk. He charges the Lakers' bench and hits anyone who comes near him. He breaks John Block's jaw, sees Rudy and hits him two more times. Willis punches out practically everyone on the team except for Jerry, who's sitting at the end of the bench in street clothes, and me. I want no part of this rampaging crazy man. I've been injured enough. Besides, I'm sort of mesmerized by the way Willis hits one Laker and then another and then *another,* each guy falling backward into the next one like a line of toppling human dominoes.

At some point, Mendy and Richie restore order and I assume Willis gets kicked out. We continue the game shorthanded, half our team bloodied, their noses broken, and we lose by three. After the game, it occurs to me that none of the Knicks came over to help Willis. I ask Dick Barnett about that. Dick shrugs. "He was doing fine by himself."

The next night, I score 45 against the Bulls, but our walking wounded bunch loses again. I'm crushed. I felt so optimistic at the start of the season, but after the Brawl, we never seem to get it together. With Jerry out and the new guys trying to find a flow and still lacking a strong center, we continue to stumble. I do have my old confidence and swagger back, moving without pain or hesitation, and I average 32 points a game.

And then, at home against those same Knicks, in the fourth quarter of a tight game, I dive onto the floor for a loose ball and a Knicks player falls on my left leg. It feels as if someone has crushed my knee with a hammer. I bite my lip to stop myself from screaming. Somehow I get to my feet, the Sports Arena bursting into applause. The next day, Bob Kerlan gives me the verdict: I haven't torn or broken anything, thankfully. I've just severely sprained my knee.

I miss three weeks. By the time I return to the lineup, in late November, joining Jerry for the first time this season, our record stands at 5–12. We never recover.

. . .

We end the 1966–67 season with a dismal 36–45 record, finishing third in the West, barely beating out the expansion Chicago Bulls. I can count the highlights of the season on one hand.

There's the All-Star Game in San Francisco. I'm once again chosen as a starter, along with Jerry, and Darrall Imhoff also makes the squad. Amazing that three of our starters make the All-Star squad and yet we have our worst season in years.

For some reason, I really enjoy this game. I play twenty minutes and score 20 points, bag five rebounds, and hand out five assists. Rick Barry, the league's leading scorer, wins the MVP award, and we upset the East, with Wilt, Oscar, and Russell, 135–123.

In February, we play the Philadelphia 76ers, sporting a 47–6 record and flanking Wilt with Luke Jackson, a bruising forward, and streak shooters such as Chet Walker, Hal Greer, and Billy Cunningham. This game, played at the Sports Arena, turns into the kind of wild and funky game we used to play back on the D.C. playground. Just the way I like it.

I dip, drive, dunk, and high-fly my way to 44 points, and we stun the 76ers, 143–133. A couple of nights later, the division-leading San Francisco Warriors and Rick Barry come to town. We crush them by 49, 129–80.

And that concludes the season's highlights.

But I'm not worried. We'll turn the season around in the playoffs.

Then, in the last game of the season, against the Bulls, Jerry breaks his hand. He's done for the season. And so are we.

But I plan to go out fighting.

We draw the San Francisco Warriors in a best-of-five series. We don't make noise in the first two games. I face double and triple teams and struggle to score. In game three, I've had enough. Rick and I put on a show. We each score 37, but he has help, too much help, and the Warriors sweep us, 3–0.

I don't really care about individual stats, as I've said, but this season I look at them to confirm that I've pretty much returned to form. I average 26.6 points a game, fourth in the league, behind Barry, Oscar, and Jerry. I also pull down 13 rebounds a game, not bad for an undersize forward playing nightly in the land of giants. And the ultimate compliment: I once again make the All-NBA First Team, with Wilt, Oscar, Jerry, and Rick.

But I feel frustrated.

I admit it. I would like to win an NBA championship. Just one. If we're to do that, something has to change.

Someone once said that the definition of insanity is doing the same thing over and over and expecting a different result.

I don't think the Lakers' management is insane. So, in order to win an NBA championship, we have to do something different.

• • •

Rick Barry jumps.

The league's scoring leader and All-Star Game MVP bolts the NBA to become the centerpiece of the new American Basketball Association's Oakland Oaks.

"Why would he do that?" someone asks me.

"Five hundred thousand reasons," I say.

That's the number I hear — $500,000. An astonishing figure.

Without a doubt, professional basketball has become *the* hot sport. The television audience keeps growing, meaning more advertising revenue. More owners want in, and the league continues to expand, this season adding two new teams: the Seattle SuperSonics and the San Diego Rockets, with more teams to be added next season. From city to city, teams outgrow their small, inadequate arenas, and new venues sprout up everywhere. Not so many years ago, we played our home games before scant crowds in the Minneapolis Armory, a dimly lit barn where the dribble of a basketball echoed eerily off the walls. This December, we'll move into Jack Kent Cooke's showplace, the Fabulous Forum, capacity 17,500 — a gaudy replica of the ancient Forum in Rome, offering luxury boxes and several restaurants. I'm told that the ushers will wear togas as they escort fans to their seats.

And now, competing with all of this, this upstart new league, the

ABA, arrives, rolling out in or close to NBA cities. The commissioner of the league, George Mikan, the former Minneapolis Lakers great, envisions an up-tempo game played with a red-white-and-blue ball. I've seen photos. The thing looks like a beach ball.

But — *$500,000?*

Rick Barry is an excellent player, but he's only been in the league for two years.

I have to ask. What kind of cash would this new ABA pay Wilt or Oscar or Bill or Jerry or — *me?*

I can't shake thoughts of my somewhat uncertain future. I will soon turn thirty-three. I know I have two or three good years left, but —

What would the ABA pay?

I don't put out any real feelers, but I do admit to some curiosity.

Somehow a feeler comes back to me. Wilt and I would each receive $600,000 *plus* ownership in a new ABA team based in Los Angeles.

I laugh.

Even if I did jump to the ABA along with Rick Barry and Wilt, whom would we play against? Where would they get players? Can you imagine the diluted talent pool? Right now, I suit up every night against the best players in the world. Besides, I would never leave the Lakers. I started with the Lakers and I plan to retire as a Laker. I wouldn't leave the NBA, even for $600,000.

If you wanted to make it a *million* . . .

Somehow the Lakers' front office hears about the feelers. Or maybe they anticipate that the new league will make me an offer I can't refuse. Either way, before the end of the summer, the team extends my contract through 1970 and gives me a nice pay bump. I will make $100,000, the top salary in the league, joining an elite club that includes Wilt, Bill, and Oscar.

• • •

The summer of 1967 does bring change — and turmoil. Sadly, our cities again burn after racial incidents. Riots and killings spread through poor sections of Tampa, Minneapolis, and Detroit. President Lyndon Johnson escalates the war in Vietnam. I thank my lucky stars that I served in the Army Reserves a couple of years before we sent thousands of troops into what seems like an ill-advised, unwinnable war. In late April, Muhammad Ali refuses to serve in the military and the World Boxing Association strips him of his title. Two months later, Ruby and I attend a dinner at the White House, guests of President Johnson. We've come as part of an American contingent to honor the king and queen of Thailand. I spend the evening enjoying the meal and the president's hospitality, answering everyone's questions politely. But my mind keeps wandering to images of rioting, Vietnam, Ali. Changes.

Back in L.A., the Lakers make changes, too.

Our general manager, Lou Mohs, passes away suddenly, and Fred Schaus, tired of coaching, steps in to replace him. We lose Walt Hazzard and John Block to the two expansion teams, but otherwise Fred makes only one significant roster change: dealing Rudy LaRusso to San Francisco. I don't say anything, but I don't agree with the move. I like Rudy as both a friend and a player, and I think San Francisco has gotten a steal. I turn out to be right about that: Rudy has his best year in 1967–68, averaging 21.8 points a game for the Warriors.

Meanwhile, we need a coach. Fred decides not to hire any of the usual suspects, career coaches who have been fired and hired several times by multiple teams, and goes with Willem "Butch" van Breda Kolff, the former coach of Princeton University and an ex-Marine.

Butch arrives with a swagger, tons of enthusiasm, and a huge personality. He doesn't merely light up a room — he lights up a room like a prison yard searchlight. Butch does nothing quietly or tentatively. He knows exactly what he wants, and if that doesn't work, he knows exactly what he wants instead. He shouts, cajoles, uses body English, and then shouts some more. He jokes, roars at his own jokes, and, after games, hits the nearest bar, drinks hard, and parties harder, with or without the team. Looking back, I'm shocked that Fred hired Butch. The two are total opposites. I think of Fred as one of the original grumps, morbidly serious, always in a bad mood. Butch shines with optimism and spirit

—you work hard and then you *play*. Early on, he shares his philosophy with the team.

"If we win, we're going out drinking. If we lose, we're going out drinking."

I like Butch from the jump.

He'll turn out to be, hands down, my favorite coach.

After getting to know us through watching us practice and studying game films, Butch institutes his "Princeton offense." In simple terms, the offense works on movement—ball movement and player movement, everybody touching the ball, passing the ball, setting screens, cutting to the basket. He believes, from what he's seen in the films, that we all dribble the ball too much. When we look at the footage, with Butch pointing at the grainy black-and-white action projected on the floppy portable screen in front of us, I see what he means. Jerry or I will dribble . . . and then dribble . . . and then dribble some more, our heads up, looking for an open man or an opportunity to drive to the hoop or shoot, but until then we dribble . . . and we dribble . . . and we dribble some more. In certain painful sequences, on our way to our worst season in L.A., the ball seems to die in the vast cavern of *us,* the two of us, disappearing into blurry, yo-yoing hands.

Butch insists that his Princeton offense will change that.

I buy into it, and so does Jerry.

It takes half a season for Butch's offense to really kick in. We get off to a slow start, partly due to another rash of injuries to Jerry. By season's end, as we enter the playoffs, Jerry will have missed thirty-one games. I will constantly test my knees, and finally come to terms with my chronically aching back. Yes: my back. I realize I'm only now mentioning my bad back. Just as when I played, I've chosen to ignore what will be diagnosed as a weird curvature in my spine, causing me persistent pain, which worsens the older I get.

Still, with all of my injuries—weak knees, aching back—I somehow miss only five games all season.

Butch prefers a small lineup, usually starting Jerry, me, Tommy Hawkins, Gail Goodrich, and Archie Clark, the second-year guard from the University of Minnesota, who picks up the scoring slack when Jerry's out. Archie averages 19.9 points a game and the league names him, along with Jerry and me, to the All-Star Game, played this year in New York before a record crowd of more than eighteen thousand. The East wins the game, as usual, 144–124, but I score 22 in front of the typically crazy crowd in Madison Square Garden, always my favorite place to play on the road.

.   .   .

Butch loves to dance. Especially after he's had a few drinks. For some reason, he seeks out places I might call dives: lots of locals, a jukebox, sometimes a bar band doing bad covers, usually sawdust on the floor. Most coaches stay away from partying with their team. Not Butch—the more Lakers around him, the better. I know some guys feel awkward drinking with the coach. I don't. I enjoy Butch's company. And watching him dance? Hysterical.

After getting a good buzz on, Butch invariably gravitates toward the biggest woman he can find and escorts her onto the dance floor. He has moves I've never seen before—moves nobody has ever seen. Often, he'll go drinking with his wife, a quiet woman who doesn't dance but seems to enjoy watching Butch do his improvised dance moves. For Butch, life is too short to brood after a loss. To be clear, Butch hates to lose as much as I do, as much as anyone does. He just refuses to allow losing to bring him down. You have to shake it off. You have to shake it *out*.

On New Year's Eve 1968, we play our first game at the Fabulous Forum, taking on the lowly San Diego Rockets. Employing the Princeton offense, we move the ball from player to player, the basketball flying around like a hot potato, and we bomb the Rockets, 147–118, with eight Lakers scoring in double digits. After the All-Star Game, we go on a tear, at one point winning fourteen of seventeen games and averaging close to 130 points a game. On February 11, we blast the Boston Celtics out of their own building, annihilating them 141–104, their worst home loss ever. After our so-so start, we finish the season 52–30, second in the West. Personally, I feel thirty-three years *young*. I average 26 points and 12 rebounds a game and make First Team All-NBA, joining Wilt, Oscar, Jerry Lucas, and Dave Bing.

• • •

March 24, 1968.

We prepare to take on the Chicago Bulls in the best-of-seven Western Division Semifinals. We come into the playoffs healthy and confident. The Philadelphia 76ers, led by Wilt, have dominated the league this year, and all the so-called experts predict they will storm through the playoffs and repeat as NBA champions.

They'll have to go through us first.

Jerry expresses essentially the same sentiment to *Sports Illustrated*, saying that our team gets along better than any team he's ever been on. He describes us as aggressive on defense and marvels at how well our team can shoot. In the article, he sounds beyond optimistic. He sounds *happy*.

We wreck the Bulls, four games to one. Bob Boozer, now starting at forward for Chicago, tries to check me, but he might as well try to guard a plane propeller. At least that's how I feel — whirling and spinning ferociously in every direction. I average close to 30 points a game in the series and put up 37 in our game five blowout, 122–99. But our defense makes the difference. We stick to the Bulls like tar. We hold them to an average of 99 points per game over the course of the series.

Next we take on the San Francisco Warriors for the Western Division Finals, beginning at home in the Forum on April 5.

On April 4, the night before the finals for both the Eastern and Western Divisions, Martin Luther King Jr. is murdered.

I frankly don't know how I will play the next night. I feel stunned, angry, and deeply sad. *Our* voice has been stilled. I feel that the NBA should postpone all playoff games. I call Bill Russell, player-coach of the Celtics. He tells me that some players want to play and several want to postpone. I don't ask which players want to play; I don't want to know.

I picture certain players on the Celtics and even guys on our team—
mostly from the South—who I doubt can understand what Dr. King
meant to me, to us. I hear comments: "He wasn't the president or anything. Why should we call off the game?"

The games go on as scheduled, but President Johnson declares Sunday, April 7, a national day of mourning, and that lifts my spirits somewhat.

Still, my entire body slumps as I approach the hardwood floor in the
Forum. We demolish the Warriors by 25, but the whole game I feel like
I'm in some kind of a dream. I play the San Francisco series in a fog. It
helps that we simply outman them. We blow them out in four straight
games, with poor Rudy LaRusso trying to check me. He can't stay with
me and I torch him in every game. After the final game, in which I score
28, we meet each other at midcourt and throw our sweaty arms around
each other.

"Sorry we swept you, Musty," I say.

We now await the winner of the Celtics–76ers seven-game war,
which, surprising the experts, goes to the Celtics in overtime in the final
game.

The press descends.

Reporters ask me how I feel. They ask how we will employ Butch's
new Princeton offense. I explain that I love the ball movement, having
everyone touch the ball. But, I say, when the game is in doubt, you have
to have the confidence to take over, to take the shot. I love our team, I
say. We have a bunch of unselfish players. But if we're behind—

I shrug and smile.

They bring up my age, my knees, the fact that I play many roles
on the team, depending on circumstances—scorer, ball handler, rebounder.

"Rebounding takes the most out of you," I admit.

One reporter points out that I have already gathered ten thousand
rebounds in my career.

Now, *that* makes me pause. Ten thousand rebounds. Ten thousand
jumps, bumps, shoves. My body recoils just thinking about it.

I'm not heartbroken that we will be facing the Celtics instead of the younger Sixers. The Celtics have added veterans Bailey Howell and Don Nelson, Bill turned thirty-three earlier this year, Sam Jones must be at least our age, and Havlicek is no kid. Bunch of old guys. They have to be worn out after that Philadelphia series. Overall, we're younger, quicker, and motivated. We'll beat them this time, our sixth NBA Finals appearance against the Boston Celtics. Sixth time is the charm. Has to be.

Except it isn't. We lose. Again.

The Celtics take us out in six games. I don't shoot well. My whole game feels off. I play too fast, or too slow, and even though I score my average, the points come harder. And Bill blocks dozens of shots, including one of mine. I get revenge in the next game. I take off at the free throw line and dunk over him. I'm not making any kind of statement; I'm just frustrated and pissed that we cannot match them in the middle. Mel Counts is a seven-foot shooting center who can't deal with Bill. I'm not complaining, and I'm not disrespecting Mel. I'm stating a fact. In order to beat Boston, we need a center. A real center.

We get him.

• • •

Early in the summer, I get a phone call from Fred Schaus.

"We have a chance to trade for Wilt."

For a moment my mind drifts and I hear only the sounds of my neighborhood—a leaf blower, a pool splash, kids' laughter.

Wilt?

Jerry and I playing with Wilt? Really?

"Elgin, you there?"

I realize I've gone silent.

"Yes . . . yeah . . . I'm here."

"What do you think? I wanted to talk to you first before I pull the trigger. I want to make sure you're good with this. You're the captain, the team leader."

"I appreciate that." I stare off and try to picture our team with Wilt in the middle. Will he fit in? Will he be content to set up high, at the free throw line, screening for Jerry's outside shots and taking out my man as I drive toward the basket? Or will he insist on playing down low, his usual spot, clogging everything up, causing gridlock in the lane? Just what L.A. needs — more gridlock. And yet —

This could be our only chance to beat Boston.

I will turn thirty-four at the beginning of this season. I don't want to say it's now or never, but it's pretty damn close. We can't stand pat. We can't beat Boston with Mel in the middle. We need Wilt. He dominates the league like no other center, and he can match up with Russell. Wilt has changed his game. Have to give him that. He has given up his desperate need to score every time he touches the ball and now concentrates on passing, rebounding, and defending. He seems to have *become* Bill Russell, or at least his version of Bill, the Bill Russell in his mind. Wilt won an NBA championship in Philly just over a year ago with this mind-set. He could do it again with me and Jerry. But Wilt and I have an uneasy history, going back to the playgrounds of D.C. We're not exactly close. I know that Bill and Wilt sometimes go to dinner together, but I wouldn't consider them close, either. I don't know if Wilt is close to anybody. Jerry himself just said that he loves this team, the guys, the chemistry. Will Wilt ruin all that?

"What do you think?" Fred says.

"You know I've been saying that we need a dominant center for years. But I never thought we could get Wilt. What would we have to give up?"

"Archie Clark, Imhoff, and a draft pick."

Hate to lose Archie. But we are talking about Wilt.

"What does Butch think?"

"Butch?" Fred hesitates. "Butch is very open to the idea."

A little too long a pause there.

"Then okay, do it," I say.

"Good."

Outside, a siren blares, jarring me.

Same sound that signals the end of a game. Or a prison riot.

. . .

Fred and Jack Kent Cooke make the trade, sending Archie, Darrall, and Jerry Chambers to Philadelphia for Wilt. At first I hear that Butch grumbled, saying something about breaking up a great team and questioning why we would even make a trade like that. Then Butch changes his position and goes way over the top to a reporter, boasting that with Wilt we will have the best team in basketball history. The press runs with that. They hand us the NBA championship in July, before Wilt even arrives.

Training camp convenes one morning in late September. I arrive early. Driving to the Loyola University gym, I make a decision. It's a new season, a new start, and, as captain, I will welcome Wilt, my new teammate, warmly. I know that the team tends to follow my lead, so I will go out of my way to make things comfortable between us. I promise myself I'll have a good attitude.

I do, for about twelve seconds.

First, he's late.

Butch, who always runs late himself, can't stand it if you're late. He paces in the locker room and bristles when Wilt finally walks in.

"This was not my idea," Butch mumbles to me as he fastens on a frosty smile and strides over to Wilt, his hand outstretched. I follow close behind. I shake hands with Wilt and then step back as the other guys form a kind of welcome line, greeting him like it's his bar mitzvah. I notice then that Wilt has gotten huge through his upper body and arms. He looks misshapen. He has the chest, shoulders, and arms of a bodybuilder, mounted on the skinny legs of a marathon runner.

"You've gotten big over the summer, man," I say after everyone finishes meeting him.

"I've been lifting weights," Wilt says. "I've also been eating healthy. Cut down on my red meat. Oh," he says, turning and speaking to the room, searching for the equipment manager, "I notice you only have Gatorade and Pepsi. I need to have 7Up. It's the only thing I drink."

"I'll take care of that," the equipment manager says from the back of the room.

"And I need grape soda," I say. "Only thing *I'll* drink. Otherwise I can't play."

I mean this as a joke. Maybe it comes out wrong. Maybe I've tried to kid around with Wilt too soon. He doesn't laugh; he glowers. A few of the guys laugh, a few don't. Everyone feels the tension. I shrug. Could be a long season.

All through training camp, I try to connect with Wilt. I invite him to play poker with us. He joins us after bragging about his card-playing prowess. Maybe my running commentary during the game throws him off, but he's a poor poker player. He loses his shirt. What's worse, he claims he forgot his wallet. He offers me and the other guys IOUs. We tell him that's not necessary—we're friends, teammates, and we know he's good for it. He never pays.

He also turns down my dinner invitation, saying he's made other plans. I offer to reschedule. He says he'll get back to me. He never does. In our four years on the Lakers together, we never go out to dinner.

It becomes obvious that Wilt doesn't want to socialize with any of us. He keeps to himself, arrives on game days later than he's supposed to, and after games he bolts as soon as he can. I rarely see him out with anyone on the team, except, eventually, Jerry. I'm not sure if Jerry asks Wilt to join him for dinner because he feels sorry for him, or if Wilt asks Jerry, and Jerry, feeling put on the spot, can't say no. I know they go out to dinner a few times. One of those times, Jerry reports to me the next day that Wilt spent the entire meal complaining. I'm sure that's true, because Wilt complains a lot. He complains about Butch's personality and his coaching, or, more accurately, as he says, Butch's inability to coach. He complains that he's always open and the guys never get him the ball. "Why should *that* guy shoot?" he scoffs. "What is he even doing in the game?"

He tells me privately that Jerry ignores him when he's open, which leads me to believe that he tells Jerry privately that I ignore him when he's open. Of course, he's not always open. He just thinks he is or says he is, even if he's blanketed by two or three guys. If I take a shot, he'll shout to me, "Didn't you see I was open?" Wilt could be sitting on the bench and if somebody in the game takes a shot, he'll say he was open.

And then we have Wilt the ladies' man. After he retires, Wilt famously boasts in one of his books that he slept with twenty thousand women. I find that claim slightly hard to believe, since I never saw him with even one woman. Another reason I doubt he slept with that many women? I never saw him take a shower. I used to call Rudy LaRusso "Musty" when he played for us. I dub Wilt "the Big Musty."

Many years after I retire, on a special occasion at the Forum, a few of us gather to take a picture — Magic, Jerry, Kareem, Wilt, and me. As we prepare to pose for the picture, Wilt starts to feel faint. Someone leads him back to the locker room so he can lie down. I go back to see him. I lean over to him and say, "Was it the twenty thousand and *one* that got you?"

• • •

Butch and Wilt butt heads from the beginning.

As I expect, Butch tells Wilt that he wants him to play the high post, setting up outside, near the free throw line. He tells Wilt specifically that he wants him out there because that will give me more room to drive.

Wilt scowls, puts his hands on his hips, and stalks Butch, finally cornering him. "How am I supposed to rebound if I'm out at the free throw line?" he says, towering over him like an awning. "Isn't that why I'm here? To rebound and block shots? I already changed my entire style of play — I don't look to score. But I belong underneath the basket."

Butch doesn't back down. "You're gonna play where I tell you to play."

This begins a running battle. Wilt complains about playing the high post — to me, to Jerry, to Butch, to anyone who will listen. In games, Wilt starts out at the free throw line, but before long he drifts back to his customary position in the low block. At that point, Butch yanks him. Wilt sits on the bench, seething, until Butch inserts him back into the game. The pattern repeats: Wilt sets up at the free throw line, then sneaks back beneath the basket, stealthily, as if nobody will notice a seven-foot, three-hundred-pound human being skulking toward a different position on the court. Everybody notices. Butch notices. And then he sits him.

Wilt rebels in other ways as well. He comes and goes as he pleases, occasionally misses practices, shows up when he feels like it. If we have a morning shootaround scheduled before a game, he will sleep in, delivering this message to the coaches: "You want me at the shootaround or the game? Pick one. Because I'm not doing both."

He really pushes it right before our season opener in Philadelphia, where Wilt has family. He comes into the dressing room before the game holding a shopping bag. He takes a seat, reaches into the bag, and pulls out a tray of fried chicken and a bucket of potato salad. He chows down, right in front of us, gnawing on fried chicken as if he hasn't eaten in a week. He doesn't offer us any of his home-cooked meal—not that any of us would stuff ourselves right before a game. At one point, Butch comes over, holding his clipboard, stops, stares at Wilt, and says, "Wilt—"

Wilt turns to him, a glob of food packed into his mouth, and says, "Mmm?" and then he belches.

Eating heavily before a meal becomes a common part of Wilt's pregame prep. As does eating at halftime. He often sends one of the ball boys out to get him a snack—nachos, burgers, a hot dog; I don't see much health food—which he washes down with a liter of 7Up. The man can eat. I decide that he likes to settle into the low post and stay there because he's too bloated to move.

• • •

And yet—I maintain a good attitude.

I do see Wilt as our antidote to Bill Russell. I appreciate what he brings to the team—rebounding, defense, and his offensive presence —and focus on how we can work together to win a championship. We seem to have all the pieces. In addition to Wilt, Fred brings in Keith Erickson, a six-foot-five defensive specialist who played at UCLA, and Johnny Egan, a steady five-eleven point guard who can back up Jerry and knock in the occasional outside shot.

I, too, have been working on my shooting. I know that Wilt, with his wide body, by screening for me even once in a while, will provide more open looks from the outside. Butch knows it, too, and suggests I concentrate on shooting during practice. He also knows how much I hate to practice, so he devises a game. He divides all of us into two teams and sets up a sort of "shooting course," spots on the floor at different angles and distances from the hoop—ten feet, fifteen feet, twenty feet. The first team to make ten shots from every spot wins. Butch, knowing me, doesn't have us play for fun. We play for stakes. I no longer remember how much or what we play for—could be Cokes, might be money. It doesn't matter: as long as I have *something* to play for, I will beat you. My team wins every time, even when the other team has Jerry.

We ease into the season tentatively, all of us, especially with Jerry, Wilt, and me trying to fit in with one another. Despite my issues with Wilt, including politically—the guy actually supports the Republican, Richard Nixon, in the upcoming presidential election over Hubert Humphrey, my choice and the choice of every other black person I know—I try not to allow anything to get in the way of our playing together. I see it partly as a question of survival. After turning thirty-four, while I don't want to admit it, I've started to feel my age. My knees ache more at the start of this season than in previous years and sometimes creak when I bend. And for the first time ever, I feel fatigued at the ends of games. Formerly the team leader in rebounds, I cede that role to Wilt. This season he will haul in 21 rebounds a game, triggering fast breaks, and giving *me* a break at the same time.

We lose our opener, badly, in Philly, win the next game, then lose a couple before we begin to figure one another out on our way to a three-game win streak. I play well those first two weeks, averaging 31 points a game. Finally hoisting the championship trophy remains an elusive vision in my mind, but I have to admit, I like our team. After a big win against the San Diego Rockets, crushing them 152–116, I tell a *Los Angeles Times* reporter, "With Wilt on our side, it makes the game much easier for me to play." I'm actually starting to believe that.

Then, at the end of October, Wilt's father dies. I don't know if it is sudden or if his dad was sick, but Wilt goes into a funk. I tell him I'm sorry about his loss. Wilt mutters thanks, but for several games he isolates himself even more. I can see he's hurting, but he seems to overcompensate, complaining more, if that's possible, criticizing the role players on the team, and escalating his constant bickering with Butch. They no longer hide their disdain for each other. They don't even try. Wilt calls Butch "the dumbest coach I've ever had" to anyone within shouting distance. About Wilt, Butch says, "Big Dipper, my ass. He's the Big Load."

Amazingly, we continue to win. Going into the All-Star Game, our record stands at 30–15, and even though Jerry again misses several games due to injuries, we continue to play better.

Baltimore hosts this year's All-Star Game, and Wilt and I represent the Lakers. I'm named a starter—Nate Thurmond of the Warriors gets the nod over Wilt—and I play one of my best games. I score 21 points, leading the West. I feel relaxed and move freely, exploding to the hoop on a couple of swooping drives. Still, the West loses, as expected, 123–112, with Oscar scoring 24 for the East and winning his third All-Star Game MVP trophy.

Wilt and I head back to Los Angeles for a few days' rest before we play Cincinnati for two games. Jerry returns, we beat the Royals both nights, and I score 27 and 34. We move on to Milwaukee to play the expansion Bucks, a team we beat easily for our third win in a row. During these three games, I notice that Butch changes his substitution pattern. He benches Wilt—a lot—replacing him with Mel Counts. In the Milwaukee game, Mel scores 26, while Wilt, spending several minutes sulking on the bench, a towel draped over his head, scores only four. I can't tell whether Wilt has suffered an injury or has just pissed off Butch.

Butch's substitution pattern continues. He'll start Wilt, then sit him for no reason I can see. One night in late January, against the Royals, Wilt, obviously wanting to make a point, ignores everybody on our team, abandons his role as a passer, rebounder, and defender, and decides to score every time he touches the ball. He doesn't ask for the ball. He demands it. He scores 60, his eyes fierce, piercing. We win 126–113. Wilt storms out of the dressing room after the game, refusing to make eye contact with anyone.

We keep winning through January even as Butch and Wilt's relationship deteriorates even more. They freeze each other out, talking only in short, snide bursts. On February 1, we play the Warriors in San Francisco in the first half of a home-and-home set, and beat them in overtime, 106–101. I score 34. I don't concentrate on Butch's substitution pattern, but I notice that Wilt again seems to sit a lot in favor of Mel, who scores 22 to Wilt's 11.

We return to L.A. to take on the Warriors again the next night. This time the game goes into three overtimes before we lose, 122–117. Nobody's talking after this game. I feel exhausted, beat up, worn out, and frustrated. I score 30, same as Mel. Wilt puts in 23 but misses eight free throws. I can't blame the loss on him. We all shot poorly from the free throw line, and Jerry again has missed the game. Still, even with the fraying relationship between Butch and Wilt, we'd won five in a row before that to push our record to 38–17. But Wilt seems on the verge of erupting. He bangs around the locker room after the game, swearing under his breath before he barges out the door.

We fly up to Seattle for our third game in three nights. On the flight up, and spilling into the locker room as we dress for the game, a nasty, pulsing tension follows the team like a fog. Then it happens. I don't recall exactly *how* it happens — I'm at least twenty feet away — but I see Butch walking down the corridor outside the locker room and Wilt lumbering toward him. I hear Butch say something in a raised voice about being sick and tired of Wilt not hustling. Then Wilt raises his voice and says, "You don't know shit about substitutions."

"Shut your mouth," Butch says.

Wilt whirls on Butch. "Nobody talks to me that way," he says, and then they charge each other — Butch, fearless, foolish, an ex–Marine sergeant, lunging at Wilt, the seven-foot, out-of-control, out-of-his-mind giant. I sprint toward them, wrap my arms around Wilt, and pull him away. Within seconds, other guys appear and shove themselves between Butch and Wilt. I press Wilt into a back wall and tell him, softly, to take a couple of deep breaths and calm down. I look Wilt in the eye and see nothing but rage. For a second, I'm positive he will come at me, too. I brace myself and again tell him to breathe. I feel his body deflate. He exhales slowly and nods. I nod back and drop my arms. He pushes past me and walks away from Butch, charging back into the locker room. Butch, restrained by two players, thrusts a finger at Wilt's back and shouts something at the locker room door.

• • •

The almost-fight between Butch and Wilt hits the papers. I'm not surprised. They went at each other in a fairly public space, and reporters must have overheard the two of them screaming. Plus, by this time players on our team have chosen sides, and I'm sure some of that drama has leaked to reporters. I know that a few players align with Wilt because they believe he will bring us that elusive NBA title, even though he complains continuously, defies Butch, and ignores team rules. They see Butch as a wild man and a drinker (he is), a screamer and chair thrower (he is), and an impetuous, spur-of-the-moment decision maker who doesn't know what he's doing (sometimes I wonder). Butch's followers don't want to rock the boat, realizing he controls their playing time. In the NBA, minutes equals money. The more you play, the more coaches rely on you, resulting in job security.

As for me, I find myself stuck in the middle for the first time since John Castellani coached the Lakers. I tolerate Wilt because I believe we are a better team with him in the lineup than we are with him sitting on the bench. I tell this to Tommy Hawkins, my friend and our union rep, a few days after the altercation in the corridor.

"The owner wants a team meeting," he says.

"That won't do any good," I say.

"Tell me about it."

"The Big Load," I say, shaking my head.

"You know what Wilt tells me if I drive to the hoop?" Tommy says. "'Don't cut through here when *I'm* in here. I'm trying to *work*.'"

"He thinks we're all in his way," I say.

We have the meeting with the owner, Jack Kent Cooke, a flamboyant figure who says something about our team being like a stock portfolio and all of us needing to invest in each other. I don't know what the hell he's talking about. I certainly don't feel as if the meeting or his metaphor accomplishes anything. Then, at Wilt's request, Tommy holds another meeting, this one for players only. Absent Jack Kent Cooke and Butch, players feel safe to air their grievances. I stay uncharacteristically quiet. I have already come to terms with the grim, inevitable truth: nothing will change unless either Butch or Wilt steps aside. By now the tension has morphed into division. As if to make my point, Wilt, essentially ignoring the whole team, drops 66 points on the hapless expansion team the Phoenix Suns. Wilt doesn't say "Fuck you" to Butch, but he may as well have.

• • •

Miraculously, we keep winning. Talent triumphs over attitude. Talent triumphs over *everything*. Even when Butch benches Wilt for Mel just to spite him, we still win. Gradually, without any discussion or debate, we simply settle into our roles. We all commit to playing defense. We smother our opponents, often holding teams to less than 100 points a game. I score when I need to, run the floor, hit the open man, and pull down my share of rebounds, more than 10 a game. Jerry, when he's even halfway healthy, pours in points. Wilt, too, reluctantly seems to embrace his role on the floor. He clogs the middle, blocks shots, and rebounds with a vengeance. He pulls down 27 boards in one game, and against the Celtics and Bill Russell, he goes insane and rips down 42 rebounds. We beat the Celtics four out of six times in the regular season. On March 7, we beat them in overtime, 105–99, and nine days later, we travel to Boston Garden and beat them by 35 points.

Slowly, subtly, I start to pace myself in games, gliding to the rim when I have a clear path to the basket, roaming outside and spotting up when I have a good shot, moving and cutting without the ball to get open, picking my spots when I want to improvise, twisting and corkscrewing myself inside for dunks or easy layups. I score 34 against Boston in that March 7 game, 10 points more than my average. Then, on Friday, March 21, before our game at home against the Hawks (who this year moved from St. Louis to Atlanta), I prepare for an event that I have tried not to think about too much.

Elgin Baylor Night.

Nice. Except that I don't like attention and I try to avoid the spotlight. Elgin Baylor Night feels a bit . . . premature. I'm not going anywhere. I'm not retiring. I'm not old. Well, I'm not *that* old. But at the insistence of the Lakers, I'm having a night.

As the evening gets closer, I hear that we've sold out the Forum. More than seventeen thousand fans have bought tickets. In addition to Ruby, Alan, and Alison, who all dress up in new outfits bought for the occasion, the Lakers fly my family in from D.C. Mother and Pops, dressed as if they're attending a Hollywood premiere, come in, as do my brothers and sisters, my nieces and nephews. I wear a gold warmup suit and pace the sideline. When I first see my whole family gathered in one place, all here to honor me, I lose it. I bow my head and turn away, trying to cover up my tears. After a moment, I pull myself together and someone leads me to a hundred-year-old chair previously owned by the president of Mexico. As I sit, to thundering applause, I wonder if this chair is supposed to be a throne signifying my success or an easy chair symbolizing retirement. I glance at my father, who points to the chair and grins at me. I picture him sitting in his chair, camped in front of the television set. Then I see myself sitting in my own chair at some point in the (hopefully) far-off future. *Every old guy has his chair,* I think.

Local dignitaries speak. The mayor. City council members. Hollywood celebrities. Members of the visiting Atlanta team shake my hand, and one by one, each of my teammates congratulates me. I receive tributes and gifts—a gold watch, a pool table, a car. Tommy Hawkins, our team spokesman, reads a short statement. He says that I am Charlie Parker with a basketball. *You never know what he's going to do, and he never does the same thing twice.* I blink, beyond touched, and then the tears come again. Tommy calls me the greatest basketball player who has ever lived. I don't hear much else, and I remember even less. I know that when the ceremony has ended, I pull myself out of the chair and wave to the crowd, everybody standing and cheering for what seems like ten minutes. As the spotlight lifts off me, I walk slowly to the sideline to prepare for the game against the Hawks.

I glance at my family. I have tried to stay close, emotionally if not geographically. I have succeeded for the most part with my mother and my sister Gladys. Columbia, so close to me when we were kids, has struggled to find her place—in the world, in life. She has fought all sorts of demons. She has closed herself off, and when she shuts herself away, I tune out. I don't do that because I am cold. I do that because I feel helpless. I love her so much that I fear the pain she is in will also do me in. So, to protect myself, I keep my distance. I love her deeply. She knows that. I hope knowing is enough.

On Elgin Baylor Night, we beat the Hawks. That's about all I remember of that game. My family cheers. The crowd gives me another standing ovation when I exit the game in the fourth quarter. As I head to the bench, I catch another glimpse of that hundred-year-old presidential chair. My throne. My easy chair. My symbol of retirement.

• • •

I don't play in our final game of the regular season, a 128–111 drubbing of the New York Knicks. As we've previously agreed, Butch gives me the night off so I can rest my knees and my increasingly sore back. We finish the season 55–27, first in the Western Division, our best record yet. Even with Wilt crowding me out from beneath the basket, and with Jerry, Wilt, and me learning how to share the basketball, I've averaged 24.8 points, 10.6 rebounds, and 5.4 assists per game, and I missed only six games all year. The league again names me to the All-NBA First

Team, with Oscar, Billy Cunningham, and two Baltimore Bullets: Earl "the Pearl" Monroe and Wes Unseld, who wins both Rookie of the Year and MVP. As we prepare to take on the San Francisco Warriors at home in the first round of the Western Division playoffs, guys on our team say that Baltimore, New York, and Philadelphia have supplanted the Celtics as the best teams in the East.

I shake my head. "I would not count out the Celtics."

"They're not that good anymore," somebody says.

"They have Russell," I say, and turn away.

Point: me.

. . .

To pretty much everyone's shock — including mine — we lose the first two games at home to the Warriors. Nate Thurmond, the Warriors' swift, athletic center, dominates Wilt. The rest of us play with no energy, no urgency. I score 12 points in the first game and lead our team with 20 in the second game. Then, warming up before game three in San Francisco, something happens. My world, as I know it, changes.

I suddenly feel tired.

Not physically tired. Not emotionally or mentally tired.

I feel . . . spiritually tired.

As if I have arrived at some sort of conclusion.

I see myself sitting on that hundred-year-old chair on Elgin Baylor Night and I wonder if that night I celebrated the player I was and not the man I am.

Bill and I talked about this once at dinner.

When it's time, how do you make a graceful exit?

It's not time, I tell myself. *I'm fine. I feel good.*

Standing in a semicircle as we warm up, I take a pass from somebody, and instead of launching up a shot, I start to dribble. I bounce the ball up and down, hardwood to waist, the ball an extension of my hand. Then, in what feels like slow motion, I wheel around and look at the guys warming up around me: Bill Hewitt, a thin, energetic rookie forward from USC, who appears more focused now than I've ever seen him. Johnny Egan, Tommy Hawkins, Mel Counts, Keith Erickson. Wilt. Lurking beneath the rim, tipping in missed shots, adjusting his head-band, lost in his own world. And Jerry. Draining jump shot after jump shot, the ball arcing, his arm flicking forward, his sheer will overpowering his brittle body, fire brimming in his cold steel eyes.

Up and down I dribble, the *whap-whap-whap* of the hardwood a faint echo, and I ask myself: *How much more?*

How much more have I got?

How much more can I give?

"Elg!" somebody calls.

I twitch, roll my shoulders, take a step back outside the corner of the key, pick up my dribble, and launch a jump shot.

I miss.

• • •

We don't lose another game to the Warriors. We hold them to an average of 90 points per game, and in game six, when we close out the series 4–2, we blow them out by 40 points. I average 12 points per game in the series.

For the Western Division Finals, we draw the Atlanta Hawks, a collection of athletes who like to run, led by Zelmo Beaty, Joe Caldwell, and Lou Hudson.

We ground the Hawks in five games, spreading the points around and again relying on our stifling defense. We take a 3–1 lead into a nationally televised Sunday afternoon game from Los Angeles. I play effortlessly, and I play above the rim. I fly back in time. I am at Spingarn. I am in Idaho, Seattle, Minneapolis, the Sports Arena. The game comes to me, and then I *take* the game from everybody else, any way I want it. Wilt lumbers out of my way and lets me drive. Jerry hits me on run-outs. I spin, snake between opponents, fake defenders out of their sneakers. I score 29 points, pull down 12 rebounds, and hand out 11 assists. The next day, Jim Murray, the *Los Angeles Times* columnist, writes, "Do you believe in ghosts?" suggesting that the Elgin Baylor of old — the ghost of me — haunted the Forum to close out the Hawks.

Yes, I think, *Jim Murray sees it, too.*

I am spiritually tired.

Once again, we will play the Celtics for the NBA championship. Boston remains standing after two gut-check series, first with Philly and then with New York. This is it.

After we dispatch the Warriors and the Hawks, pundits and Vegas oddsmakers establish us as overwhelming favorites to win the championship. The Celtics, everyone says, look tired. Sam Jones has announced that he will retire after this series, and player-coach Bill Russell can't be too far behind. Boston has brought in veterans Bailey Howell and Don Nelson to back up John Havlicek, the man with the nonstop motor, and added Emmette Bryant, a spark-plug guard with a deadly jump shot. Some people will call the 1969 NBA Finals the greatest ever.

Having the better regular season record, we receive home court advantage, playing the first two games in Los Angeles. In game one, Jerry takes over. He scorches the Celtics for 53 points. I add 24 but mainly stay out of his way. Despite his outburst, the game goes back and forth until the very end, and we hang on to beat them 120–118, in front of a sold-out, delirious home crowd.

Game two. Jerry stays hot and pours in 41, but tonight I swirl, shoot, and leap to 32 points, powering past Havlicek, who fouls out, and Don Nelson, who ravages me like some kind of offensive lineman. Butch rests me strategically, making sure I find a second wind for the fourth quarter. I dominate in the fourth, scoring at will. I've been hearing a lot of talk about Boston "stealing one" in L.A. I go on patrol, playing cop and home court protector. Boston will not *steal* anything on my watch. We win 118–112. We fly to Boston for game three, the guys feeling confident, loose. That is, all the guys except Wilt, Jerry, and me. Jerry looks miserable. He's been suffering from migraines, but I think he's in pain because, although we've taken the first two games, we haven't won anything. Not yet. Wilt looks glum. I think he, more than anyone, feels the pressure. Jack Kent Cooke and Fred brought him in to win a championship and to win it now. This year has seen the death of his father, dissension with his coach, and uneasiness with his teammates. He can't feel comfortable, I know that.

We come out flat in game three, get our bearings, then Jerry goes cold and I can't find the hoop. I feel as if I'm trying to throw a ping-pong ball into a shot glass. We go down by 17 points, then come roaring back and cut the margin to three, before they step on our throats and knock us off, 111–105.

It's all right—I didn't expect to sweep the Celtics. But I want to win game four. I don't want to leave here with a split. If we can take game four, we have them. I know it: we have them.

Game four is a mess. Neither team can hold on to the ball. I play one of the worst games of my life. I manage to score two baskets and go one for six from the foul line, for a total of five points. Still, Jerry scores 40 and we stay neck and neck with the Celtics. Then, late in the fourth quarter, with both a one-point lead and the ball, I try to seal the game. I break free from Nelson and race downcourt as Wilt winds up and heaves a long pass in my direction. Unfortunately, the ball sails and wobbles like a dying duck. I reach for the ball, snatch it, and tiptoe along the sideline, making sure I stay inbounds, at the same time feeling somebody pushing me in the back.

The referee blows his whistle.

He says I've stepped on the baseline. He bends down and points to it repeatedly like a crazed mime.

I jump so high in protest, I nearly catapult over our bench.

"You're out of bounds," the ref says, gesturing wildly, the crowd erupting as he waves his hands violently.

"I was *in!*" I shout. "If I did go out, I got pushed!"

The ref tucks the ball into his side and walks away from me. I start to go after him but stop. I can't change his mind.

"It's a bad call," I say under my breath. "You made the wrong call."

As I turn to pick up my guy and play defense, I'm hoping that this call doesn't turn the series around. One call can't change a whole series —can it?

Yes, it can.

With one second left in the game, Sam Jones hoists up a prayer from just inside midcourt. The ball looks like it has no chance to go in — it actually looks like he's passing to someone under the basket — but the ball eases over the front of the rim and drops in.

We lose, 89–88.

One call. One bad call.

We go back to Los Angeles, the series tied two games apiece.

The comforts of home — that's what we need, and that's what we get. We run Boston out of our gym, 117–104, behind Jerry, who scores 39, and Wilt, who pulls down 31 rebounds. I score eight points and feel like a spectator, wondering if maybe I should have bought a ticket. Then, with about two minutes left in the game, Jerry pulls up lame. He grabs the back of his leg, his eyes pinched in pain. Hamstring. He limps to the bench.

We fly to Boston for game six. I come out determined to carry us. I want to win this series here, now, in Boston Garden.

But it's not to be. I score 26, and Jerry, hobbling, matches me, but Boston jumps out to a quick lead and we can't catch them. They beat us 99–90.

We head back to L.A. for game seven.

I know, in the deepest part of me, that this is it.

I know that, while this may not be literally my last NBA game, this is my NBA Final.

• • •

May 5, 1969.

Jack Kent Cooke brings in the USC marching band and stuffs the Forum's rafters with thousands of balloons that will drop the moment the horn sounds after we win our first NBA championship.

Not one person on our team thinks this is a good idea.

Some see the balloons as a bad omen.

Jerry worries that the balloons will provide incentive for the Celtics.

I don't love balloons or marching bands. But I don't think about them. I just want to play this game.

An overflow crowd fills the Forum, and from the very beginning I feel their nervousness. They cheer, of course, but mostly I sense a level of anxiety. We play the Celtics even until early in the third quarter, when they start to pull away. The fourth quarter begins and we fall further back, trailing Boston by 16. Jerry, dribbling quickly, breaks the Boston press and passes the ball to Johnny Egan, who finds me on the left side of the key. I fake my defender, dribble toward the middle, cross the free throw line, and float in for a leaning half-hook, half-scoop shot. *Bang*. We narrow the gap to 14. Russell sprints downcourt ahead of everybody, takes a long pass from Em Bryant, and slams in a deuce. Boston back up by 16.

The crowd buzzes, waiting, hoping. The clock ticks.

Jerry makes one of two free throws, the Celtics set up their offense, and Bailey Howell drills in a line-drive jump shot.

Celtics by 17.

We trade baskets, we trade misses, then Jerry hits a jumper to cut the lead to 15.

The Celtics slow down their offense, Bryant misses a long jumper as the twenty-four-second clock expires, the ball gets batted around, and I snag the rebound. I dribble across midcourt and snap a pass to Jerry on the left side. He gets bumped as he tries a jump shot.

He hits one of two foul shots.

Celtics by 14, 8:28 left in the game.

The Celtics, a running team, walk the ball upcourt. This is exactly what they shouldn't do. This is exactly what we need: for the Celtics to alter their usual crisp rhythm. Again late in the shot clock, Jerry blocks a Havlicek jump shot from behind and Wilt grabs his 20th rebound. We jog upcourt. I pass to Keith, who hands the ball to Jerry at the side of the basket. He nails a ten-foot jumper.

Celtics by 12, 7:58 left.

Boston calls timeout.

Back on the floor, Don Nelson drives on me. I slap his arm before he can score. He misses the foul shot. Jerry misses a jumper, but Tommy rebounds, then has a lay-in blocked by Russell. Wilt gets the rebound and is fouled trying to put up a shot. He makes one of two free throws.

Celtics by 11, 7:25 left.

A scrum under the Celtics' basket. Russell grabs the rebound and tosses it out to Havlicek, who drops the ball, but Tommy is called for fouling him. Livid, Tommy screams and stomps his feet. Jerry pulls him away. Havlicek makes the foul shot.

Celtics by 12, 7:15 left.

We sprint upcourt, Sam Jones sagging off Jerry to cover Wilt. Jerry dribbles across the middle. Sam leaves Wilt and comes at Jerry. Jerry stops and fakes. Sam bites and runs into Jerry. The official blows his whistle, calling a foul on Sam.

That's six on Sam. He has fouled out of his last game. He walks toward his bench, his head down, and the Forum crowd erupts, standing and cheering for Sam Jones's last appearance in the NBA.

Jerry makes the free throw.

Celtics by 11, 7:05 left.

The Celtics miss their next shot. Jerry takes his time dribbling upcourt. I hang out on the right wing and sneak behind a Wilt screen. Jerry hits me in stride. I rise up for a jump shot. Swish.

Celtics by 9.

Havlicek hits a jumper. Jerry answers.

Celtics by 9, 6:15 left.

Havlicek misses again, Wilt grabs the rebound, but when he comes down, he lands awkwardly and tweaks his knee. He shouts in pain and grabs his leg. He can't seem to move. Players maneuver around him. Another Celtic shoots and misses. Wilt grabs this rebound, too, despite the look of agony on his face, and calls timeout. Hoggy runs over, shaking a can of numbing spray. He sprays Wilt's knee with the stuff. Wilt grimaces. He can't seem to put pressure on his leg. Hoggy says something to Wilt. Wilt shakes his head. Hoggy jogs back to the bench, and Wilt stays in the game.

Jerry goes up for a short jumper and a Celtic fouls him. Jerry steps to the line for two shots as Wilt, his head lowered, waves to the sideline and limps to the bench. Butch sends in Mel Counts to replace him.

I glance at Bill Russell. Exhausted, he leans over and grips his shorts. He watches Wilt walk toward the bench. A strange look comes over Bill. I can't quite identify what I see in Bill's eyes — relief? Concern?

No. I see — disgust.

Bill seems revolted that Wilt would take himself out of this game. We are playing for the basketball championship of the *world*. Wounded or not, you play. You gut it out. You find a way. And then, for the first time ever during a game, Bill catches my eye. I read this expression perfectly.

Elgin, you play wounded all the time, his eyes say. *You always play in pain.*
You've played on one leg.
You've played on no legs.
You play.

The game resumes, with Mel Counts now in the post.

Jerry makes his two foul shots.

Celtics by 7, 5:16 left.

The Celtics attack the basket. Em Bryant shoots, misses. Mel taps the rebound to Johnny Egan, who immediately finds Jerry downcourt. He stops, shoots — swish.

Celtics by 5, 4:50 left.

Jerry has 40 points.

The Celtics miss another shot. Keith Erickson grabs the rebound, passes to Jerry, who hits me for a ten-foot bank shot. Three Celtics surround me and derail the shot. Erickson flies in for the offensive rebound and cleans out Don Nelson. The ref blows his whistle. Keith waves his hand in frustration. As the crowd roars, screaming encouragement, we walk toward the Celtics' end of the court for Nelson to shoot his free throw.

Sitting on the bench, Wilt fidgets and tells Tommy Hawkins, sitting next to him, that he's ready to go back into the game. Tommy leans over to Butch and relays the message, saying that Wilt feels good to go.

"We're doing fine without him," Butch says.

Stunned, Wilt ignores this and starts to get up.

Butch whips toward him. "Sit your big ass *down!*" he screams.

At the foul line, Don Nelson squats, pauses, and hoists his strange, shot-put free throw.

He misses.

I yank down the rebound.

I flip the ball to Jerry, who scats upcourt.

Jerry dashes into the corner on the left side, where two Celtics converge on him. He flings the ball to me just as one of the Celtics whacks Jerry across the forearm, drawing a whistle. Jerry steps to the foul line and cans his two free throws.

Celtics by 3, 4:11 left.

I look at our bench, expecting to see Wilt getting ready to check into the game. Instead I see him sitting, fuming, a towel draped over his head.

Em Bryant brings the ball up. He tosses it to Larry Siegfried, who hands it to Havlicek. John dribbles, penetrates, stops, and shoots. He misses. Mel grabs the rebound. We head toward our basket. The crowd noise builds, surges. Usually I can drown out crowd noise, but not this time, not today. The crowd has reached a level of hysteria. Johnny Egan gets the ball back to Mel, who stops at the foul line to shoot but gets called for traveling. Counts redeems himself on the next play, blocking an Em Bryant reverse layup. Egan rushes downcourt and takes a fifteen-foot jumper but misses. Boston takes the ball up and Siegfried dribbles into the lane, popping an elbow into Mel's chest on his way. He's whistled for an offensive foul, and Butch calls a timeout, with the Forum full-house noise level redlined, deafening.

We trail by 3 with 3:21 left.

We walk to our bench, and every player on the team stands to meet us.

Except Wilt. He sits slumped, staring past us.

Butch says something. Not sure if he's calling a play or screaming encouragement. I can't hear him.

We head back out on the floor and I look over at Wilt. He raises his head and looks at me.

I read the message in his eyes. I read it loud and clear. *If you lose this, don't blame me.*

We take the ball in. Jerry passes to Egan, who passes to me. Back to Egan, who finds Mel Counts, standing alone on the free throw line. Mel shoots.

He hits.

Celtics by 1, 3:01 left.

We head the other way. Russell asks for the ball inside, to work on Counts. Bill faces up to the basket for a two-foot shot—and misses. Mel grabs the rebound. He flips to Jerry, who dribbles into our end. We can take the lead. Down by 17 points less than eight minutes ago, we can take the *lead*. We have the momentum. The crowd is going berserk. Jerry and I make eye contact. He knows I want the ball. He knows I can give us the lead, and then . . . I can't say it . . . I can't allow myself to *think* it . . .

I streak to an open spot on the right side. Jerry whips the ball to Egan, who gets it to me. I jump and launch a ten-foot jump shot. So easy. So perfect. The ball arcs off my fingertips—

And drifts an inch too far, pinging off the back rim.

Havlicek and Erickson battle for the rebound and the whistle blows. The refs call Keith for a foul.

Havlicek steps to the foul line, his team leading by one.

An excellent foul shooter, he will give the Celtics a two-point lead.

He misses.

Celtics by 1, 2:30 left.

Jerry waves his arm, telling all of us to clear out. He wants to go one-on-one and give us the lead. He wants to take this game. We move out of his way. He drives.

Siegfried reaches in and taps the ball loose. It skips out of Jerry's hand and bounces into Boston's hands.

The Celtics fly downcourt.

They lose the ball.

1:56 left.

Jerry dribbles quickly, hits Egan, who passes to Erickson in the corner, who sees me beneath the basket. Alone. I'm alone. I will not miss this shot. I will not miss this shot. *I will not miss this shot.*

Erickson passes me the ball.

Russell zooms in front of me and intercepts the pass.

Celtics by 1, 1:33 to go.

Siegfried drives the length of the court and heads into the right corner, where two Lakers trap him. Siegfried somehow gets the ball to Havlicek, who dribbles toward the hoop, shadowed by Erickson. Keith reaches around Havlicek and pokes the ball away. The loose ball skitters toward the free throw line and right to Don Nelson. Nelson picks up the ball — a gift — and shoots his strange shot-put jump shot.

Too hard.

The ball slams into the back rim and bounces high into the air — straight up, at least three feet into the air —

And then — incredibly, defying the laws of physics — the ball drops straight back down and nestles into the net.

Celtics by 3, 1:15 left.

I know we have more than a minute left, but that crazy, impossible shot feels like a kick in the gut. I know it's over.

Jerry misses a jumper. We stop the Celtics and get the ball back.

Forty-six seconds to go.

Jerry passes it off to Egan, who gets the ball in to Counts. Determined, Mel drives and attempts a weird inside shot — I can't really tell what he's doing or why — but he goes right at Bill Russell, and Bill, in a gesture of unmitigated defiance, swats the shot out of Mel's hand and to a teammate.

If Wilt had taken the same shot . . . If Butch had put Wilt back in . . .

I can't go there. I can't.

We foul Siegfried, who hits two foul shots, Boston steals it, and we foul again. Havlicek hits a free throw, the lead is back to six, and then, with twelve seconds left in the game, Egan puts up a meaningless jump shot.

In that moment, I go back in time twenty years.

I am fifteen years old, in a pickup game on a D.C. playground.

I am Rabbit.

Among a group of older boys, of men, I time my jump and knife in for the rebound, beating everybody, then put the ball back in. Nobody can stop me.

Now, beneath a ceiling packed with thousands of pathetic balloons that will never be released, before I can shoot, a delirious Celtic grabs me by the arm.

I step to the foul line for two shots.

I look up at the scoreboard.

Celtics 108, Lakers 102.

Seven seconds left.

The crowd begins to file out.

I make the first foul shot.

I squint. I roll my shoulders. I look down. I breathe. I eye the basket. I shoot.

I barely remember my second shot going in.

With no time on the clock, Johnny Egan drives in for an uncontested layup to make the final score Boston 108, Los Angeles 106.

We lose to the Celtics again — the seventh time we've lost to them in the NBA Finals.

Do I think Wilt should have been in that game? Would it have mattered? Would we have won the championship? I don't know.

I know that he made only four of thirteen foul shots and we lost by two points.

If he had hit just three more foul shots, we would have won. Nobody has mentioned that. Still — I don't know.

I take a last look at the stupid balloons hanging overhead, a blue-and-gold helium ceiling, and, feeling numb, I walk off the court and into the locker room. I have left every ounce of my spirit and my soul on that basketball floor.

I'm not ready to go.

I am not *ready*.

I sit in the locker room, my head bent, avoiding everybody's eyes. Guys walk by on their way to the shower. Some mumble incomprehensible words, sentences I can't decipher. Some just rest their hands on my shoulder or my arm for a moment as they pass. I grunt in acknowledgment. I lift my head and see Butch in the corner, his face a white mask. He looks dead. He says nothing. I know he's thinking about a drink — several drinks — and I wonder if he'll ask me to join him. I might. I don't. I have a sudden sense that I will never play for him again. I avoid Wilt. I don't know where he went. I can't look at Jerry.

The locker room clears out, and still I sit.

Butch's hollow voice rings through the empty room. "We were doing better with Mel."

Did he say just that? Or did I dream it?

But then I hear another voice, a voice from years ago, the chirpy voice of Ray Felix: "We'll get 'em tomorrow."

We'll get 'em next year.

We won't.

We won't ever get 'em.

10

FINALLY

AFTER GAME SEVEN OF THE 1969 NBA FINALS, BILL RUSSELL announces his retirement. Shortly after that, Butch van Breda Kolff announces his resignation. He resurfaces sometime later as coach of the Detroit Pistons. I'll hear from someone reliable that he actually signed his deal with the Pistons *during* the 1969 NBA Finals. I'll never believe that Butch kept Wilt on the bench for any reasons other than that he honestly felt we were playing better without him, and that he couldn't stand him. Fred scours successful college programs for a new coach and hires Joe Mullaney, a quiet, button-down guy from Providence College who gives the impression that he once worked for the FBI or the CIA because he did.

I turn thirty-five a month before the start of the 1969–70 season, my twelfth in the NBA. My knees have never felt worse. I play only 54 games. I still average 24 points and 10 rebounds a game, and again I start the NBA All-Star Game. I play every game as if it's my last. I love the guys. I love the game. I love the wars.

Wilt misses most of the season with a leg injury.

Jerry enjoys his best year as a pro, averaging 31 points per game and missing only eight games.

Wilt comes back in time for the playoffs. We again reach the NBA Finals, this time against the New York Knicks, a machine of a basketball team, led by Walt Frazier and a hobbled Willis Reed. We fight them hard, but they blow us out in game seven at Madison Square Garden.

Two games into the 1970–71 season, I sever my Achilles tendon.

I miss the rest of the season. I spend the next year rehabbing my body and my life. As the months go by, I face the inevitable. I know that I have come to the end. I just can't say the words. I can't accept the end, I guess. And so I don't accept it. I convince myself and everyone around me that I can come back next year, at age thirty-seven. I can play one more year. Somehow. I can.

But I don't really believe it.

The Lakers fire Joe Mullaney and hire former Boston Celtics All-Star guard Bill Sharman, a disciplinarian who loves to practice, instituting two-a-days. I arrive at training camp fired up. I slip easily in with this group of guys, among them Wilt, Jerry, Gail Goodrich (who will lead the team in scoring), Leroy Ellis (who returns to back up Wilt), Happy Hairston, Flynn Robinson, and a young forward from Columbia University whom Sharman loves, Jim McMillian.

We open the season with four games on the road. Sharman starts me at forward, along with Happy Hairston, but in every game, he takes me out in the first quarter in favor of McMillian. We win each game easily. In the fourth game on the road, against the Atlanta Hawks, Sharman pulls me two minutes into the first quarter. McMillian goes on to score 39 in a one-sided win, while I, playing a few meaningless minutes, score seven.

After our home opener, a 113–106 loss to the Chicago Bulls, in which I score 19 and for the first time this season feel halfway like myself, I get a message that Bill wants to see me in his office. I have a strange feeling about the meeting. I put off seeing him until after our next game, on the road in Houston. I twist my knee in that game and spend almost all of it on the bench. Finally, after we lose on Sunday night, Halloween, to Golden State, I go to see Sharman the following Thursday in his office.

Sharman doesn't waste any time. He barely looks at me. "Beginning tomorrow I'm going to start Jim McMillian," he says.

I say nothing.

"He's earned the chance to start," Sharman says.

"Okay," I say, because that's really all I can say, all I want to say.

"What do you think?"

What do I *think?*

I think —

I want to go for Sharman's throat and flip his desk over.

I want to bolt.

I want to retrace time.

I think —

So this is how it ends?

Being replaced by a kid.

I sigh.

And then I actually chuckle.

The end.

I try it out in my head.

The end.

The. End.

Hell, I already came to the end, in 1969, the last time we lost to the Celtics. I knew it then. I just couldn't move on it then. Man, it would have been something if Bill Russell and I had walked out of Dodge together, a couple of gunslingers hanging up our holsters forever.

"What do I think?" I say, sitting military straight in the chair opposite Sharman. "What do *you* think?"

He doesn't expect that.

"I'm not sure I —"

"If I decide to *retire.*"

Not a question. A statement.

"Retire? Well. I guess . . . I mean, you have two more years on your contract. We'll pay what we owe you, of course, and we'll work something out, maybe a consult —"

"Yeah," I say, standing, not wanting to hear any more. "That's what I'm going to do. I'm going to retire."

. . .

So, nine games into the 1971–72 season, I retire.

Just like that.

No fanfare. No farewell tour. No spectacular sendoff. I just leave. I interpret my meeting with Sharman as being shown the door, so I go. I get out.

The Lakers, I'm told, will hold a ceremony to retire my number, 22. Okay. Nice.

Mal Florence writes my farewell story in the *Los Angeles Times,* with the headline BAYLOR ENDS FABULOUS CAREER AFTER 13 YEARS.

In the article, Mal says, "No man has contributed more to the success and the popularity of the Lakers since they moved to Los Angeles from Minneapolis in 1960 than Baylor."

For the record, I retire as the third-leading scorer in NBA history, with 23,149 points, trailing only Wilt and Oscar. I finish with the third-highest scoring average per game, 27.4. I also retire as the fifth-leading rebounder of all time, with 11,463 boards. Not bad for an undersize six-foot-five forward.

But, as Ray Charles said, the secret to life is timing.

The 1971–72 Los Angeles Lakers win thirty-three consecutive games, a record that still stands and I doubt will ever be broken. Jerry and Wilt — and Jim McMillian — finally accomplish what eluded me for thirteen years, winning an NBA championship by beating the Knicks easily. To this day, Jerry says his biggest regret is that he couldn't share that championship with me.

• • •

I loved playing basketball.

You know what? That statement doesn't come close to how I felt —how I feel. Playing basketball was *everything* to me — my escape, my release, my therapy, my joy, my life, my definition. Who I am.

But I didn't love watching basketball for a living, or commentating about basketball, or analyzing basketball, or making basketball decisions as a general manager. I did all of these things after I retired from playing.

In October 1974, Pops passed away. When I first heard the news, I felt numb. Then I felt sad. Then I felt angry—not at him, but *for* him, for all he went through as a man, and as a black man. Then I allowed any residual ill feelings I had toward him to fade and I forgave him, and then I tried to forgive myself.

In January 1975, after being separated for more than a year, Ruby and I divorced. There's not much more to say about that except that over the years we grew apart. As I said, I probably married her for the wrong reasons. I don't think she ever loved living in Los Angeles, one of the issues that became a wedge between us, a problem, because I always considered L.A. home. Ruby moved back to D.C. with our kids, Alan and Alison. I took care of them financially. I love my kids and always will.

I took a job with the expansion New Orleans Jazz for the 1974–75 season. I started out as an assistant coach, and became the head coach a couple of years later. I went to New Orleans for a change of scenery and to reunite with Butch van Breda Kolff, long gone from the Pistons and now coaching the Jazz, and Hot Rod Hundley, the team's radio and TV broadcaster.

In hindsight, deciding to move to one of America's premier party towns with two of America's premier party animals may have been a mistake. Butch, his reputation preceding him, never met a New Orleans bar he didn't like. Hot Rod, as always, enjoyed a nightcap or two, or three.

On the court, the Jazz struggled, but it wasn't really Butch's fault. They had no *players* until they brought in Pete Maravich. Unfortunately, Butch and Pete never got along. Butch had this idea that he should rein in one of the most prolific collegiate scorers of all time. Butch and Pete clashed, and in late 1976 Butch got fired. The team offered me the job. I took it. Reluctantly.

I unleashed Pete. The kid had averaged something like 40 points a game in college, so it made sense not to hold him back. He flourished. But we still didn't win very much. A lot of the other players resented him, especially the black players, saying that Pete got special treatment because he was white. I have no idea where that came from, because, in case no one noticed, I happened to be black. In the end, I didn't love coaching. Pete got hurt, the team didn't win, and I got let go in 1979.

I also didn't love New Orleans. I missed L.A. But something happened to me there before I left that upended my world.

· · ·

In April 1975, to quote former president Bill Clinton, I met a girl.

Elaine Cunningham.

A mutual friend introduced us.

"You *have* to meet this woman," he told me.

I didn't really want to meet anybody, but I agreed to speak to her on the phone. He called Elaine and told her about me. She said she wasn't interested.

"An athlete?" she said. "I have a thing about athletes."

"What?"

"I don't like them. They're full of themselves. They're boring and totally narcissistic."

"He's different. Are you sitting down? He's — Elgin Baylor."

"Who?" Elaine said.

I called her on the phone. We spoke for two hours. I completely lost track of time. So did Elaine. And as we spoke that first time, I knew. I had met the One.

We got married on September 17, 1977, the day after my forty-third birthday. We settled back in Los Angeles, and in 1982 our daughter, Krystle — the other love of my life — was born.

From the moment I told Elaine "I do" until this moment, no matter what loss I experience, what challenge I face, what pain I feel, I take a deep breath and remind myself, "You have Elaine."

And she has me.

· · ·

Donald Sterling.

There, I said it.

I debated whether to mention him at all.

Sportswriter Bill Simmons suggested that my years as general manager of the L.A. Clippers under owner Donald Sterling diminished everything I'd accomplished as a basketball player. I don't know if that's true, but it is possible. I worked for Donald for twenty-two years, starting in 1986. Although we didn't share an office suite—we didn't even work in the same building—I got to see him up close. I would call him mean, bordering on abusive, especially to his employees, incredibly cheap, and, yes, racist. I offer a few examples of each, but only a few, because that's all I can stomach.

Donald rarely confronted people in person. He preferred to scream at his employees on the phone and then hang up on them. One afternoon, I saw how cold and heartless he could be. Donald asked me to come to his office for a meeting. I drove across town from my office, walked into his reception area, and took a seat next to a well-dressed woman, a manager of one of the buildings Donald owned. The woman seemed anxious. She kept fidgeting and looking at her watch. After a while, she went over to Donald's secretary and said, "I'm really sorry to do this, but I've been waiting to see Mr. Sterling for over an hour. Would you tell him that I had to leave to pick up my daughter? I'll be back as soon as I can."

A little while later, the door to Donald's office opened and Donald appeared. He asked his secretary to send in the building manager.

"She had to leave to pick up her daughter," his secretary said. "She'll be back as soon as she can."

"When she comes back," Donald said, "tell her she's fired."

• • •

It didn't take long for me to find out how cheap Donald was. After only a few weeks as general manager, I learned that he refused to pay players salaries that were commensurate with what the rest of the teams in the league paid. As the years went on, I often got criticized for some of the draft picks I made and some of the deals I didn't make. I wish I had kept track of all the phone calls I received from college coaches and agents who told me not to draft or sign a certain player because he refused to play for the Clippers. Half the time, I felt I was scrambling to put a full roster together.

We had a terrible time hiring coaches for the same reason — they didn't want to work for Donald. At one point, I convinced my good friend John Thompson, the extremely successful Georgetown University coach, who also grew up in D.C., to meet with Donald and me for the vacant Clippers coaching job. To be honest, John was reluctant to consider the job, but he agreed to talk to Donald mainly because we're such good friends.

The meeting went badly from the start. Donald started by going off on a tangent, ranting about issues that didn't relate to coaching or the Clippers. Then he abruptly stopped talking, stared at John, and said, in a snide tone, "So tell me, what have you done?"

John lost it.

"What have I done?" he shouted. "What the fuck have you done?"

Then John got hot. He spewed out a rant of his own, peppered with profanities, finally rising out of his chair to his full six-foot-ten-inch height, towering over Donald and glowering at him. He turned away, muttering to me under his breath, "I'd never work for that crazy son of a bitch," and stormed out of the office.

After John left, Donald just shrugged and said, "I don't like college coaches."

Another time, I suggested to Donald that we move Jim Brewer, our top assistant coach, into the head coaching position. Donald seemed resistant to the idea, but I persisted and Donald agreed to meet with Jim and me.

This meeting turned out to be nothing more than a formality and a waste of time. Donald asked Jim a few questions, all of which he answered politely and thoroughly. Then, obviously having heard enough, Donald waved toward the door.

"Thanks for coming in," he said. "But we're going to go in another direction."

"Are you sure, Donald?" I said. "Jim's been an assistant for five years——"

"I prefer white coaches. Black players are more intimidated by a white coach."

I stared at him. "That's a plantation mentality," I said.

"Players respect a white coach," he said.

Meeting over.

In addition to being too cheap to bring in top free agents, Donald undermined good players on our team when their contracts were coming up for renewal. He would suddenly start maligning them to me, complaining, "This guy is not a good player," or "This player can't shoot. I don't like him at all. He's terrible."

After I left the Clippers, I heard a story about one of the team's best players, point guard Baron Davis. Baron had become increasingly rattled at home games because someone in the stands constantly heckled him. Baron, an intelligent and sensitive guy, said that hearing this nonstop trash talk threw him off his game, especially when it seemed directed only at him. He felt he was being targeted. He'd be dribbling upcourt, setting up the offense, and he would hear that voice screaming at him from the front row, "You stink! You can't play! Why are you even on the team?"

It turned out the heckler was Donald.

• • •

Finally: Donald Sterling, racist.

Let's go to the audiotape.

April 25, 2014.

Everybody asked me.

Was it him?

Was that his voice?

Oh, it was him. Absolutely. No doubt about it.

"Don't bring black people to my games," Donald said on the tape, to a woman named V. Stiviano. Later, when talking about black players on the Clippers, he said, "I support them, and give them food, and clothes, and cars, and houses."

Yes, that was Donald.

In 2009, after the Clippers had removed me from my position in the front office and I sued Donald for wrongful termination, I said publicly that Donald Sterling had a plantation mentality. Now the entire world knew what I was talking about. It was right there on that tape. But what he said went beyond even a plantation mentality. To me, he sounded like a slave owner.

"Do you think he's a racist?" Anderson Cooper of CNN asked me during the only nationally televised interview I granted on the subject.

I stared at Anderson Cooper. I couldn't help myself. His question snagged me and stuck in my mind. The words felt like a wound.

Is Donald Sterling a racist?

"Of course he is," I said, feeling myself blink. "There's no doubt in my mind now. At the time, I thought he was, and now there is no doubt."

That infamous tape. The tape heard round the world. Donald Sterling talking intimately to this woman who was his girlfriend. I guess. I don't really know. Through the years, I saw Donald in the company of many different women. I don't know who they were, but he would bring women—and men, too, for that matter—into the locker room with him after games and point to the players as they went in and out of the showers. He would say, "Look at those beautiful black bodies."

The players complained. One of them called me and said, "Is there any way you could stop the owner from bringing people into the locker room while we're changing?"

I asked Donald to stop doing this. I asked repeatedly. Eventually he stopped.

I'm sure Donald knew he was crossing a line. He knew what he was doing. He just wanted to see how far he could go. At the time, friends of mine would ask me to describe Donald, his personality, his character, his behavior. I would struggle to find the right word. I think the closest I came was "different." Donald is different. Edging toward strange. I've never met anyone like him. The things he did and said ranged from unusual to outrageous. And then racist. Not at first. I worked for him for a couple of years before I heard him say anything racist. Then this happened:

In 1988, the Clippers had the first pick in the NBA draft. As general manager, I recommended we take Danny Manning, from the NCAA champion Kansas Jayhawks — a consensus All-American, the Naismith College Player of the Year, and winner of the John R. Wooden Award as the best player in the country. Everybody thought we should take Danny, me included, but even so, I told Donald that you never know what will happen when a college player — even an elite one — gets to the next level. At his request, I gave Donald a ton of research — charts, stats, and tapes of Danny Manning. He agreed with me. *Draft Danny.*

So I did.

That was the easy part.

Then came the hard part.

Signing Danny.

Donald had a reputation for lowballing players when it came to paying them. Again, Donald was cheap. I knew it, agents knew it, players knew it, everybody knew it.

After we drafted Danny, Donald invited him, his dad, his agent, and me to his house for an evening meeting. Donald answered the door in his bathrobe. I didn't catch the look on Danny's face or his dad's face, but I could feel the vibe in the room — discomfort bordering on shock. We followed Donald into his living room and we all sat down. After some small talk, Danny's agent decided to break the ice and bring up Danny's contract. The agent threw out a figure. I don't remember the number, but I'm sure it was a first salvo, probably a little inflated, trying to get the negotiation going. Donald thought for a second and countered—with a ridiculously low offer.

The agent looked a little stunned and said, quietly, "Well, you know, I don't think we can do that."

Donald looked stung.

"That's a lot of money for a poor black kid," he said.

The room went silent and cold.

Two thoughts collided in my mind.

First: Did Donald really say that?

Second: Does he know how racist that is?

I felt the rage rise into my throat. But rather than explode or lash out, I shut down. I swallowed the rage. I'm sure anyone could tell how I felt. But I maintained control, as pained and as hurt as I felt, for myself and for Danny.

As I've described, Donald had said crazy things before. He'd said hurtful things, too—not to me, but to other coworkers. I was always shocked. But when he blurted that to Danny Manning? Blatant racism, I thought. Did he mean it? Was he trying to shock us? Or did he say it just to draw attention? I knew that Donald Sterling loved attention.

All this swirled through my mind in probably less than two seconds. I snapped out of it when I caught a blur of movement in my periphery as Danny Manning streaked by me, his face locked in determination. He blazed through the room and stormed out of Donald Sterling's house. He closed the door behind him. He didn't slam it—he had too much class for that—but he shut the door firmly. He made a statement.

I looked over at Donald.

He seemed to be frowning, wearing an expression I couldn't read.

Confusion?

Concern?

Or calculation?

Was what he said some kind of tactic he was using so he would have to pay Danny Manning only the absolute minimum?

I had no idea. I could never tell where Donald was coming from.

I went after Danny. I was too late.

By the time I got outside, he had left. He got into his car and drove off.

I somehow smoothed things over with Danny, and eventually he signed with the Clippers and played for us for five and a half years. He liked his teammates. He liked L.A. I can tell you he didn't like Donald. He did well as a Clipper. He made the All-Star team and became a solid pro. Unfortunately, his rookie year he played only twenty-six games. He tore his ACL, and that ended his season. Over the years, so many players would injure themselves playing for the Clippers that fans said the moment a player signed with us, he became cursed. People started referring to a "Clipper Curse."

Danny and I came to be friends. We talked basketball and life in the NBA, and we spent time on the basketball court working on moves together.

We never talked about the incident at Donald Sterling's house.

Neither of us brought it up.

Why?

Too unbelievable.

And too hurtful.

Twenty-six years later, Donald Sterling, the man who craved attention at any cost, found himself in the center of a seemingly eternal white-hot spotlight. Someone had leaked that tape to TMZ, and the world—at least the basketball world—had exploded. Social media erupted with players expressing their shock and outrage. LeBron and Magic, among others, called Donald out. The Clippers players, in the heat of the playoffs, protested Donald's racist remarks by lining up on the sideline at Staples Center before their next playoff game with their jerseys worn inside out. A few days after the tape went viral, NBA commissioner Adam Silver appeared on national television and fined Donald Sterling $2.5 million (the maximum amount possible), banned him from the NBA for life, and strongly recommended that the other twenty-nine owners hold a vote to remove him as owner of the Clippers, in order to allow the team to be sold, which they did. Not long after, Shelly Sterling, Donald's wife, on behalf of the Sterling Family Trust, sold the team to former Microsoft CEO Steve Ballmer for $2 billion, ending one of the strangest and saddest chapters in NBA history.

· · ·

I spent twenty-two years working for Donald Sterling. Why did I stay so long? Why didn't I leave?

For one thing, I couldn't. I knew that if I tried to find another general manager's position, I would be blamed for the Clippers' losing record. Since Donald refused to pay players what they were worth, my hands were tied when it came to convincing top free agents to sign with the Clippers. The team's failure would be on me. Also, during the 1990s and early 2000s, I didn't see many NBA teams hiring black general managers. If I wanted to stay employed in basketball, I had to stay with the Clippers, as dysfunctional as that operation became. I didn't see a choice.

But the main reason I stuck it out is that I'm tremendously competitive. I wanted to win so badly. I wanted to prove to Donald, to the basketball world, to the rest of the NBA — every owner, administrator, coach, and player who scrutinized the Clippers with fascination, horror, and disbelief — that I could win despite everything being stacked against me. I wanted to win despite Donald.

Now, looking back . . .

Someone asked me recently, "Elgin, what would you have done differently?"

I don't have to think twice about that.

Everything.

I would've done everything differently.

If I knew what I was getting into — if I knew what kind of person Donald Sterling was — I never would have taken that job.

· · ·

I block them out.

The losses.

And sometimes I can't. Or I won't.

My brothers passed away—Kerman in 2006, and John in 2015. Their deaths hit me hard. My mother's passing crushed me.

Mother died on October 14, 2000. She had been sick for some time. I don't remember who called to tell me the news. I remember it was during the day. I remember sunlight filtering through the blinds in our large, open kitchen. When I heard the news, the phone slipped out of my hand. I slowly crept upstairs and got into bed. It sounds crazy, but that's what I did. I got into bed and tried to sleep off the news, tried to sleep off the hurt. I never could. I never did.

And then—many years before that, on January 12, 1979, my sister Columbia passed away in a hospital in D.C. My niece La Juan called a day or so later. La Juan told me that the doctor had taken her aside after Columbia passed.

"Who's Elgin?" the doctor asked.

"Elgin," La Juan said, confused. "Elgin is my uncle. My mom's brother. Why?"

"That's the last thing your mom said. Her last words. 'Elgin,' she said. 'I want to see Elgin.'"

I couldn't sleep for weeks. I would lie in bed, night after night, trying to conjure up Columbia's face. Staring at the ceiling the way I did after so many losses, I replayed my sister's life and mine. I replayed the loss.

What could I have done differently? What could I have done better? Columbia, what could I have done?

• • •

I sit at the counter in my kitchen. Elaine stands behind me, her hand resting on my back. Together we watch black-and-white footage of a basketball game on a computer screen. Together we watch—me.

In this footage, I watch a young Elgin Baylor.

Number 22.

In this footage, Elgin Baylor soars. He wears a Lakers away uniform and flies toward the hoop, the basketball nestled in his palm, tucked next to his ear. Pre-Jordan. Pre–Dr. J. He seems twice the size of everyone else on the court. His image takes up almost the entire frame. Other players arrive in the corner of the frame now. Even though they are in the middle of a game, they all stare up at him, even the players on his team. They watch him, their mouths open, their faces tinged with a trace of fear and wonder. Some players actually cower.

Elgin himself wears no facial expression. His body says everything. *I rule. I dominate. I fly.* The year is 1960, and this is the first time these players have seen Superman.

We watch the footage for twenty minutes. At one point, Elaine lowers her lips to my ear and whispers, "How does it feel watching yourself?"

I feel the tears fall. I look straight ahead. Elaine rubs my back slowly, in a circle. "It's all right, Elgin," she says. "It's all right."

• • •

This has been my story.

It is an American story.

It has deep roots in racism. I have known struggle and pain, and, even worse, dismissal. But ultimately, I see my story as one of survival and triumph. I fought battles — against bigotry, against cowardice — and I tried to rise above them. I didn't always succeed. But I always tried.

I never wanted to fight. I just wanted to be myself, to play the game I love so much, and when the time came and my body gave out and I couldn't play that game anymore, I wanted to leave quietly, with dignity. That's all.

In life, you lose. You can't help it. I do believe you win more. I have learned through my now eighty-three years that what matters most is not what you accomplish or how much you accumulate, but how you spend your time, and with whom. If you asked me for advice, I would say: Find love.

And then I would say: Hold on.

ACKNOWLEDGMENTS

From Elgin Baylor

First, I want to thank God for making me the way He did. I was blessed with the physique and the ability to play the way I did, a gift that provided me such great joy and satisfaction, which I was able to share with my fans and which gave me the means to provide a good living for my family. I thank God for His presence in my life and for saving me from serious and life-threatening accidents and injuries. Through life's challenges, I have had the prayers, support, and guidance of my pastors, Dan Hicks, Jim Tolle, Ricky Temple, and Tim Clark. They have been an invaluable resource to Elaine and me.

I have always gravitated to the women in my life.

I begin—and end—with Elaine.

You are my everything. You are the loving, driving force in every moment of every day. You know I always put team first; Elaine, we are an unbeatable team. I've never won a championship—except when I met you. You made this book happen.

I also want to thank a few other extraordinary women. My mother, Uzziel Baylor, for giving me life. Along with my sisters, Gladys and Columbia, you loved, encouraged, and supported me, had faith in me when I was just a sweaty "Rabbit." I want to thank my daughter Krystle for being a source of encouragement and support, willing to do whatever was needed whenever it was needed. I also must single out my niece La

Juan Graham, daughter of my beloved sister Columbia, who served as our tour guide in D.C., our family archivist, and a reservoir of invaluable information throughout this project.

And a big thank-you to the other members of my family. My father, John Baylor, who, along with my mother, worked tirelessly to provide food, shelter, clothing, and many intangibles that kids don't value until they are adults. I am grateful to my brothers, John and Kerman, who taught, teased, and taunted me just enough to keep me constantly working harder to get better. And thank you to my son, Alan, my daughter Alison, and Ruby, for your cooperation while we were in D.C.

I want to thank my close friends throughout my life, those with whom I grew up in the neighborhood, in high school, and in college: Gary "the Bandit" Mays, Clarence Hanford, Warren "W.W." Williams, Bill Wright, and R.C. Owens, along with Lloyd Murphy and Francis Saunders, who accompanied me to Seattle University. And there are so many more. I can't name you all, but you know who you are and how grateful I am for your friendship.

I wish I could thank all of my coaches, teammates, and opponents by name, but listing each of you would fill another book. Please know that I do acknowledge you and thank you. You are only as good as the players you play with and against, and every one of you made me better. I would like to mention a few, starting with, of course, the great Jerry West. We came so close, so many times, didn't we?

My great teammates from my years in Minneapolis and Los Angeles: Tommy Hawkins, Larry Foust, Vern Mikkelsen, Dick Garmaker, Bobby "Slick" Leonard, Hot Rod Hundley, Jim Krebs, Ed Fleming, Boo Ellis, Steve Hamilton, Frank Selvy, Rudy LaRusso, Ray Felix, Leroy Ellis, Gene Wiley, Dick Barnett, Darrall Imhoff, Mel Counts, Archie Clark, Hank Finkel, John Block, Johnny Egan, Don Nelson, Gail Goodrich, Walt Hazzard, Jim Barnes, Jerry Chambers, Keith Erickson, and Happy Hairston.

Among my opponents, I have to start with my two greatest rivals: the incomparable Bill Russell and the best coach I ever went up against,

Arnold "Red" Auerbach. And in no particular order, these friends and foes alike: Oscar Robertson, Lenny Wilkins, Guy Rodgers, Bob Boozer, Earl Lloyd, Willis Reed, Gus Johnson, Willie Naulls, Satch Sanders, K.C. Jones, Sam Jones, Tom Heinsohn, John Havlicek, Rick Barry, Bob Pettit, Cliff Hagan, Bob Cousy, Wilt Chamberlain, George Yardley, Walt Bellamy, Wayne Embry, Nate Thurmond, Zelmo Beaty, Bill Bridges, Bailey Howell, Clyde Lovellette, and, back in the day, Jim Wexler, Bill McCaffrey, and Gene Shue.

I also want to acknowledge all of the great Lakers who followed me, the ones who carried the torch, including Kareem Abdul-Jabbar, Kobe Bryant, Magic Johnson, Shaquille O'Neal, and James Worthy.

I thank the great people at the College of Idaho, who offered me an opportunity for a college education when other schools didn't, and the incredible folks at Seattle University, who allowed me to transfer there to continue my education. I thank all of my teammates in college, including my teammates on Westside Ford, especially Don Ogorek, Charlie Brown, Jim Frizzell, Johnny O'Brien, Eddie O'Brien, and Jim Harney, who shared so much of his time and his memories when we visited in Seattle.

I especially want to acknowledge John Castellani, who was my coach both at Seattle University and on the Lakers. John, now in his nineties, remains as forthcoming, funny, and energetic as ever. Thank you for your time. I cherish our relationship.

Bill Hogan, you are a true friend. Thank you for everything in Seattle, in particular the hours and hours you drove us around the city, even though I feared for my life every time you took the wheel.

Thanks to our trainer, Frank O'Neill, and to my good friend and one of our first team doctors, Chuck Aronberg. When it comes to doctors, I have special, loving memories of Bob Kerlan, my dear friend, a remarkable doctor, and a better human being. And I will always remember with deep affection the Voice of the Lakers, Chick Hearn. Our time together was precious.

Way back when, before the Internet, the reporters who covered the Lakers were part of our team. Thank you to Frank Deford, Jim Murray,

Mal Florence, and everyone else who covered us and wrote so positively about me.

In my twenty-two years working for the Los Angeles Clippers, I made so many good friends. I appreciate your friendship during those years. I am so happy the Clippers are now experiencing the success we waited for through so many years.

Hugh Dodson, you made it happen. You make so many things happen. Thank you.

Thanks to Anthony Mattero, my literary agent. You found us the perfect home.

At Houghton Mifflin Harcourt, thank you to Susan Canavan, editor supreme, Jenny Xu, Lisa Glover, Rebecca Springer, and Will Palmer.

Thanks to Jeanie Buss, Linda Rambis, and everyone at the Lakers. I appreciate all you do.

I thank all my close friends who support me and Elaine every day, especially Dan Hicks, Erwin and Andie Hesz, Bill and Marcia Withers, and Corky Hale and Mike Stoller.

Thank you to all the hardy fans in Minneapolis who stuck with us through thick, thin, and evil winters.

Thank you to all of the fans in Los Angeles. You mean the world to me. To everyone I missed and should've thanked, I apologize.

Finally, to Alan:

I never intended to write a book. Elaine convinced me I had a story to tell, but telling it is another matter. It took me years to agree to open up. Then, somewhat unexpectedly, you won my trust. You enabled me to reveal myself and share my life's journey. Our idea has now become a reality, and I hope my story will be encouraging to others. We are forever grateful for the work you've done and the warm relationship we've developed. Saying just thank you is so inadequate for what Elaine and I want to express.

A last word.

Jerry, as you know by now, I got my statue.

Thank you, Lakers.

You see, it really does take a village!

From Alan Eisenstock

This book, from conception to publication, took four years.

The journey I went on with Elgin and Elaine Baylor took me places I had never gone emotionally and creatively. Every step was profound, exhilarating, and joyous.

The first day I met the Baylors at their home, Elgin led me into the family room off their large and open kitchen and settled into his high-back chair. I took a seat on the couch at Elgin's right elbow while Elaine sat across from me in an armchair matching Elgin's. We talked and talked —for hours. When it got to be late afternoon, Elaine went into the kitchen and casually began to whip up some guacamole and a few other snacks. Elgin followed her and opened a bottle of white wine. We returned to the family room, toasted, ate, killed the bottle of wine, and talked some more. I didn't want to leave.

After that day, I visited the Baylor home often. I would sit on the couch and set up my tape recorder as Elgin sat in his familiar chair. Elaine would join us for some of our conversations, but most of the time Elgin and I would talk alone. At the end of two or three hours, Elaine would arrive, and soon we'd be having snacks and sometimes dinner—and always wine, which I would now bring, our ritual. And we would talk for hours. During the NBA playoffs, we'd watch a game. Without fail, Elgin would clear our plates and immediately begin washing the dishes.

"I cannot stand a dirty dish," he said.

Elgin and Elaine, I cherish every moment we spent together. I'm honored and humbled that you chose me to help you tell your story. But now that we're done and your book is out in the world, I feel a certain loss. I miss our time "working" together; I guess because it never felt like work. Here's to much more time together—more conversation, good times, and glasses of wine. We began as collaborators; I now consider you dear friends.

Elgin, thank you for your kindness, generosity, and good humor, your warm heart, and your honesty. Thank you for allowing yourself to dig deep. I know it was sometimes painful, but thank you for allowing your emotions to flow. We laughed a lot, and, yes, we cried.

Elaine, all I can say is thank you for everything. I marvel at your in-telligence, thoughtfulness, thoroughness, tenacity, passion, energy, and strength. You are an inspiration. This book began as an exciting project but became a true labor of love. There is really no other way to say this: there would be no book without you.

We did have a lot of help.

Anthony Mattero—brother in arms, partner in crime, and daily co-conspirator—who steered me to Elgin and found the perfect pub-lisher for *Hang Time*. You keep me sane and off the streets. And a special thanks to Alex Rice and everyone else at the Foundry.

Hugh Dodson, who steered Elgin to me, kept us on course, and is doing so much to take us home.

Joanna Parson, Janis Spidle, and Jessica Stahl, extraordinary ladies of letters.

My gratitude to all the members of Elgin's family, with special thanks to La Juan Graham, Elgin's niece and his sister Columbia's daughter, our tour guide both in D.C. and through many parts of Elgin's life.

Gladys Garrett, Elgin's other sister, thank you for your hospitality in Harpers Ferry and for sharing your memories. Thanks also to Elgin's nephew Algin Garrett.

Bill Hogan, Jim Varney, Johnny O'Brien, John Castellani—a great resource and raconteur—and everyone else we met and spoke to in Se-attle.

And special thanks to:

Krystle Baylor.

Jerry West.

David Ritz.

The entire Houghton Mifflin Harcourt team, starting with Susan Canavan, whose strong vision, impeccable taste, and good cheer kept me—kept all of us—going. Also thanks to everyone else on the HMH team, including but not limited to Susan's intrepid assistant Jenny Xu, Lisa Glover, Rebecca Springer, Chloe Foster, and Will Palmer.

Madeline and Phil Schwarzman, Susan Pomerantz and George Weinberger, Susan Baskin and Richard Gerwitz, Kathy Montgomery and Jeff Chester, Linda Nussbaum, Ed Feinstein, Gary Meisel, Bob Vickrey.

Jim Eisenstock, Jay Eisenstock, Susan Alon, Loretta Barrabee, Lorraine, Linda, Diane, Alan, Chris, Ben, and Nate.

Jonah, Kiva, and Randy, you make my day, every day.

Z., G.G., and S., thank you forever.

Finally, thanks to my everything — Bobbie.

APPENDIX

ELGIN BAYLOR CAREER STATISTICS

KEY

APG = Assists per game

AST = Assists

FG = Field goals made

FG% = Field goal percentage

FGA = Field goals attempted

FPG = Fouls per game

FT = Free throws made

FT% = Free throw percentage

FTA = Free throws attempted

G = Games played

GS = Games started

MP = Minutes played

PF = Personal fouls

PPG = Points scored per game

PTS = Points

REB = Rebounds

RPG = Rebounds per game

TM = Team played for

PER GAME (NBA)

SEASON	AGE	TM	G	MP	FG	FGA	FG%	FT	FTA	FT%	RPG	APG	FPG	PPG
1958–59	24	Mpls.	70	40.8	8.6	21.2	.408	7.6	9.8	.777	15.0	4.1	3.9	24.9
1959–60	25	Mpls.	70	41.0	10.8	25.4	.424	8.1	11.0	.732	16.4	3.5	3.3	29.6
1960–61	26	L.A.	73	42.9	12.8	29.7	.430	9.3	11.8	.783	19.8	5.1	3.8	34.8
1961–62	27	L.A.	48	44.4	14.2	33.1	.428	9.9	13.1	.754	18.6	4.6	3.2	38.3
1962–63	28	L.A.	80	42.1	12.9	28.4	.453	8.3	9.9	.837	14.3	4.8	2.8	34.0
1963–64	29	L.A.	78	40.6	9.7	22.8	.425	6.0	7.5	.804	12.0	4.4	3.0	25.4
1964–65	30	L.A.	74	41.3	10.3	25.7	.401	6.5	8.2	.792	12.8	3.8	3.2	27.1
1965–66	31	L.A.	65	30.4	6.4	15.9	.401	3.8	5.2	.739	9.6	3.4	2.4	16.6
1966–67	32	L.A.	70	38.7	10.2	23.7	.429	6.3	7.7	.813	12.8	3.1	3.0	26.6
1967–68	33	L.A.	77	39.3	9.8	22.2	.443	6.3	8.1	.786	12.2	4.6	3.0	26.0
1968–69	34	L.A.	76	40.3	9.6	21.5	.447	5.5	7.5	.743	10.6	5.4	2.7	24.8
1969–70	35	L.A.	54	41.0	9.5	19.5	.486	5.1	6.6	.773	10.4	5.4	2.4	24.0
1970–71	36	L.A.	2	28.5	4.0	9.5	.421	2.0	3.0	.667	5.5	1.0	3.0	10.0
1971–72	37	L.A.	9	26.6	4.7	10.8	.433	2.4	3.0	.815	6.3	2.0	2.2	11.8
Career			846	40.0	10.3	23.8	.431	6.8	8.7	.780	13.5	4.3	3.1	27.4

TOTALS (NBA)

SEASON	AGE	TM	G	MP	FG	FGA	FG%	FT	FTA	FT%	REB	AST	PF	PTS
1958–59	24	Mpls.	70	2,855	605	1,482	.408	532	685	.777	1,050	287	270	1,742
1959–60	25	Mpls.	70	2,873	755	1,781	.424	564	770	.732	1,150	243	234	2,074
1960–61	26	L.A.	73	3,133	931	2,166	.430	676	863	.783	1,447	371	279	2,538
1961–62	27	L.A.	48	2,129	680	1,588	.428	476	631	.754	892	222	155	1,836
1962–63	28	L.A.	80	3,370	1,029	2,273	.453	661	790	.837	1,146	386	226	2,719
1963–64	29	L.A.	78	3,164	756	1,778	.425	471	586	.804	936	347	235	1,983
1964–65	30	L.A.	74	3,056	763	1,903	.401	483	610	.792	950	280	235	2,009
1965–66	31	L.A.	65	1,975	415	1,034	.401	249	337	.739	621	224	157	1,079
1966–67	32	L.A.	70	2,706	711	1,658	.429	440	541	.813	898	215	211	1,862
1967–68	33	L.A.	77	3,029	757	1,709	.443	488	621	.786	941	355	232	2,002
1968–69	34	L.A.	76	3,064	730	1,632	.447	421	567	.743	805	408	204	1,881
1969–70	35	L.A.	54	2,213	511	1,051	.486	276	357	.773	559	292	132	1,298
1970–71	36	L.A.	2	57	8	19	.421	4	6	.667	11	2	6	20
1971–72	37	L.A.	9	239	42	97	.433	22	27	.815	57	18	20	106
Career			846	33,863	8,693	20,171	.431	5,763	7,391	.780	11,463	3,650	2,596	23,149

NBA PLAYOFFS: PER GAME

SEASON	AGE	TM	G	MP	FG	FGA	FG%	FT	FTA	FT%	RPG	APG	FPG	PPG
1958–59	24	Mpls.	13	42.8	9.4	23.3	.403	6.7	8.7	.770	12.0	3.3	4.0	25.5
1959–60	25	Mpls.	9	45.3	12.3	26.0	.474	8.8	10.4	.840	14.1	3.4	4.2	33.4
1960–61	26	L.A.	12	45.0	14.2	30.2	.470	9.8	11.8	.824	15.3	4.6	3.7	38.1
1961–62	27	L.A.	13	43.9	14.3	32.7	.438	10.0	12.9	.774	17.7	3.6	3.5	38.6
1962–63	28	L.A.	13	43.2	12.3	27.8	.442	8.0	9.7	.825	13.6	4.5	3.7	32.6
1963–64	29	L.A.	5	44.2	9.0	23.8	.378	6.2	8.0	.775	11.6	5.6	3.4	24.2
1964–65	30	I.A.	1	5.0	0.0	2.0	.000	0.0	0.0	.000	0.0	1.0	0.0	0.0
1965–66	31	L.A.	14	41.9	10.4	23.4	.442	6.1	7.5	.810	14.1	3.7	2.7	26.8
1966–67	32	L.A.	3	40.3	9.3	25.3	.368	5.0	6.7	.750	13.0	3.0	2.0	23.7
1967–68	33	L.A.	15	42.2	11.7	25.1	.468	5.1	7.5	.679	14.5	4.0	2.7	28.5
1968–69	34	L.A.	18	35.6	5.9	15.4	.385	3.5	5.6	.630	9.2	4.1	3.1	15.4
1969–70	35	L.A.	18	37.1	7.7	16.4	.466	3.3	4.5	.741	9.6	4.6	2.8	18.7
Career			134	41.1	10.4	23.6	.439	6.3	8.2	.769	12.9	4.0	3.2	27.0

NBA PLAYOFFS: TOTALS

SEASON	AGE	TM	G	MP	FG	FGA	FG%	FT	FTA	FT%	REB	AST	PF	PTS
1958–59	24	Mpls.	13	556	122	303	.403	87	113	.770	156	43	52	331
1959–60	25	Mpls.	9	408	111	234	.474	79	94	.840	127	31	38	301
1960–61	26	L.A.	12	540	170	362	.470	117	142	.824	183	55	44	457
1961–62	27	L.A.	13	571	186	425	.438	130	168	.774	230	47	45	502
1962–63	28	L.A.	13	562	160	362	.442	104	126	.825	177	58	48	424
1963–64	29	L.A.	5	221	45	119	.378	31	40	.775	58	28	17	121
1964–65	30	L.A.	1	5	0	2	.000	0	0	.000	0	1	0	0
1965–66	31	L.A.	14	586	145	328	.442	85	105	.810	197	52	38	375
1966–67	32	L.A.	3	121	28	76	.368	15	20	.750	39	9	6	71
1967–68	33	L.A.	15	633	176	376	.468	76	112	.679	218	60	41	428
1968–69	34	L.A.	18	640	107	278	.385	63	100	.630	166	74	56	277
1969–70	35	L.A.	18	667	138	296	.466	60	81	.741	173	83	50	336
Career			134	5,510	1,388	3,161	.439	847	1,101	.769	1,724	541	435	3,623

NBA ALL-STAR GAMES

SEASON	AGE	TM	G	GS	MP	FG	FGA	FG%	FT	FTA	FT%	REB	AST	PF	PTS
1958–59	24	Mpls.	1	1	32	10	20	.500	4	5	.800	11	1	3	24
1959–60	25	Mpls.	1	1	28	10	18	.556	5	7	.714	13	3	4	25
1960–61	26	L.A.	1	1	27	3	11	.273	9	10	.900	10	4	5	15
1961–62	27	L.A.	1	1	37	10	23	.435	12	14	.857	9	4	2	32
1962–63	28	L.A.	1	1	36	4	15	.267	9	13	.692	14	7	0	17
1963–64	29	L.A.	1	1	29	5	15	.333	5	11	.455	8	5	1	15
1964–65	30	L.A.	1	1	27	5	13	.385	8	8	1.000	7	0	4	18
1966–67	32	L.A.	1	1	20	8	14	.571	4	4	1.000	5	5	2	20
1967–68	33	L.A.	1	1	27	8	13	.615	6	7	.857	6	1	5	22
1968–69	34	L.A.	1	1	32	5	13	.385	11	12	.917	9	5	2	21
1969–70	35	L.A.	1	1	26	2	9	.222	5	7	.714	7	3	3	9
Career			11	11	321	70	164	.427	78	98	.796	99	38	31	218

COLLEGE

SEASON	AGE	COLLEGE	G	FG	FGA	FG%	FT	FTA	FT%	RPG	REB	PPG	PTS
1954–55	20	College of Idaho	26	332	651	.510	150	232	.647	18.9	492	31.3	814
1956–57	22	Seattle U.	25	271	555	.488	201	251	.801	20.3	508	29.7	743
1957–58	23	Seattle U.	29	353	697	.506	237	308	.769	19.3	559	32.5	943
Career			80	956	1,903	.502	588	791	.743	19.5	1,559	31.3	2,500

AWARDS AND HONORS

Ten-time First Team All-NBA (1958–59 through 1964–65,
    1966–67 through 1968–69)

Eleven-time NBA All-Star (1959–65, 1967–70)

1959 NBA All-Star Game Most Valuable Player

1958–59 NBA Rookie of the Year

1958–59 NBA Sporting News Rookie of the Year

1956–57 NCAA AP All-American (2nd)

1957–58 NCAA AP All-American (1st)

1958 NCAA Final Four Most Outstanding Player

INDEX